The Future of Rural Development

# THE FUTURE OF RURAL DEVELOPMENT

## Between the Adjustment of the Project Approach and Sectoral Programme Design

HANS GSÄNGER

FRANK CASS • LONDON

Published in association with the
German Development Institute, Berlin

*First published in 1994 in Great Britain by*
FRANK CASS & CO. LTD.
Gainsborough House, Gainsborough Road,
London E11 1RS, England

*and in the United States of America by*
FRANK CASS
c/o International Specialized Book Services, Inc.
5804 N.E. Hassalo Street
Portland, Oregon 97213-3644

Transferred to Digital Printing 2004



British Library Cataloguing in Publication Data

Gsänger, Hans
Future of Rural Development: Between the
Adjustment of the Project Approach
and Sectoral Programme Design. - (GDI Book
Series; Vol. 2)
I. Title II. Series
307.1

ISBN 0-7146-4104-9

Library of Congress Cataloguing-in-Publication Data

Gsänger, Hans.
The future of rural development: between the adjustment of
the project approach and sectoral programme design / Hans Gsänger
p. cm.
"Published in association with the German Development Institute,
Berlin."
Includes bibliographical references.
ISBN 0-7146-4104-9 (pbk.): £15.20
1. Rural Development. I. Title
HN49.C6G79 1993
307.1'412--dc20 93-30519
CIP

# Contents

## Summary and Conclusions

Both bilateral and multilateral donors attach considerable importance to development cooperation that seeks to achieve a lasting and self-sustaining improvement in the living conditions of the poor majority of the developing countries' rural population. It is generally agreed that only wide-ranging agricultural and rural development will eliminate the mass poverty which is particularly prevalent in the rural areas of Africa, Asia and Latin America. However, the controversy over both the means to this end and the resources to be used continues, although some meeting of the minds has recently been observed.

The current debate is characterized by the following pairs of opposites: accelerated economic growth versus general social development; effective promotion of local and regional projects versus country-wide programmes at sectoral and macro level; complex, integrated projects versus sectoral programmes geared to specific target groups; the strengthening of government organizational structures versus promotion of the self-organization of the beneficiaries; the supply orientation of public services versus demand orientation; blueprint planning versus open planning processes. This debate dominates the search for an appropriate combination of different policies and instruments for poverty-oriented rural development. This study has been made against the background of the current debate, to which it also seeks to make a contribution.

The study is divided into two main parts and eight chapters. The first part discusses the high and low points in the history of integrated rural development over the past 40 years (Chapters 1 to 4) and concludes with an attempt at a provisional appraisal (Chapter 5). The second part (Chapter 6 to 8) outlines the requirements to be met by rural development policy and development cooperation if poor rural population groups are to benefit under the conditions of the 1990s.

The study draws the following conclusions:

1. Rural development, meaning a multisectoral task having as its object the achievement of both macroeconomically adequate growth rates in agricultural production and equitably distributed social development of the rural population, has always played a relatively important role in national and international development cooperation over the past 40 years. The variously propagated promotional policies and project approaches have, however, undergone fairly rapid change; as a rule, they have been a rather blurred mirror image of the prevailing paradigms of development theory and policy.

2. The main themes of the international debate on agricultural and rural development, if divided into idealized stages each of a decade, have advanced from the modernization of traditional agricultural through the promotion of an increase in agricultural productivity to the alleviation of rural poverty and finally to the strengthening of self-help capacities.

3. Integrated approaches to promotion which seek to take account of the complexity of rural production and life by adopting coordinated, multiple intervention strategies emerged at a fairly early stage. They range from Community Development through the minimum package programmes for small farmers to approaches to the target-group-oriented satisfaction of basic needs and area development. The various basic types appeared in national guises and donor-specific forms.

4. The actual proportions of the productive and social programme ingredients in rural development have often given rise to impassioned debates among professionals. While the protagonists of growth orientation in rural development saw a socio-structural approach a priori as a costly aberration, the structuralists rightly pointed out that poverty was on the increase despite economic growth and that fundamental structural reforms were needed.

5. For a time the calls for the transfer of effective and appropriate technology and know-how and the strategic role played by

domestic research and agricultural extension services in accelerated agricultural development encouraged sectoral programmes that administered top-down innovation processes. The stroke of luck that was post-colonial agricultural development, the Green Revolution that emerged at a time when population growth was accelerating, led to unprecedented in creases in yields, especially in the countries of South and South-East Asia. However, it also helped to create fresh social differentiation, put the normally government-run agri cultural and rural service systems to the test, for which they were not always a match, and in irrigation farming gave rise to a new dimension of direct and indirect environmental problems, which are only now becoming visible globally.

6. The high and low points of rural development have followed closely one upon another. The overriding objective of developing approaches that can be replicated has been achieved in only a few cases. Although many individual projects have been very successful, they have often failed to have the desired wide impact and sustainability. Few countries have so formulated their agricultural development policies as to leave sufficient scope for both agricultural growth and social development. For their part the donors have pursued a cooperation policy which has focused heavily on projects and paid too little attention to the coherence of sectoral policies. The view that a supply-oriented strategy of providing services, know-how and inputs could set in motion sustainable development processes proved to be a serious mistake. This strategy failed in two respects: on the one hand, it reduced those concerned to functional objects; on the other, it expected government to bear fiscal burdens which were intolerable in most countries. Consumers came to expect public services as a matter of course, which led to budget deficits that were difficult to reduce.

7. From the experience gained in 40 years of efforts to achieve sustainable rural development that overcomes poverty a number of general conclusions can de drawn:

a) Economic growth is essential, but not enough on its own to limit rural poverty effectively. In many countries growth of agricultural production is hampered by structural and institutional constraints:

- appropriate agricultural technologies for typical small farms are available in only fragmentary form;
- agricultural and social services are still largely beyond the reach of the rural poor;
- almost everywhere agricultural and land reforms have failed to progress beyond the initial stages;
- agricultural market and price policies have been used to benefit urban industrial development for far too long;
- as rural-urban links are rudimentary, the opportunities for very small farmers and the landless to earn incomes from non-agricultural activities are limited;
- the absence of financial systems for specific target groups prevents the gradual integration of poor groups into the economy.

b) Satisfactory social development of the rural community calls for sufficient scope in policy and society to permit and promote self-organization, self-initiative and self-help:

- unless the people are fully involved in political decision-making and in the dialogue on development, it will be difficult to make development measures sustainable;
- needs-oriented and participatory planning for rural development with a wide impact succeeds only where the apparatus of state is sufficiently decentralized and administration is transparent;

- the structural foundations for the necessary dialogue on development are laid only when the intermediate level (third sector) between government and people has been developed and strengthened;

- the limited degree of social security and the absence of basic social services are among the main causes of mass poverty, the blocking of man's creative forces and high birth rates.

c) Rural development projects have had a limited impact in the past; the causes are to be found in their conception, the strategy for their implementation and the environment in which they have been implemented:

- integrated planning seems essential if the closely linked causes of poverty and backwardness are to be appropriately combated;

- the institutional and administrative structure in most countries suggests the wisdom of sectorally coordinated implementation;

- open planning processes are more likely to trigger self-sustaining development processes than economically stringent blueprint planning;

- the formative scope of development projects is usually very limited, since the necessary link with sectoral policies is not forged;

- macro and sectoral policies are often geared to modernization models that do not leave sufficient scope for decentralized, locally appropriate rural development.

8. Effective action to overcome mass rural poverty is one of the major tasks facing international development cooperation. As development cooperation is no longer overshadowed by stereo type thinking now that the East-West conflict is over, the

dialogue on development policies and patterns can be conducted with the poor countries more openly and with a better chance of succeeding. The initiation of reform policies in such areas as land law, technology, financial systems, primary education and health care is essential if there is to be effective orientation towards poverty in rural development. Both the conception and the implementation of such reform policies call for managerial competence and administrative capacities, with which many countries are not adequately equipped. This deficiency can be overcome by development cooperation under the structural adjustment programmes. To improve the prospects of rural development succeeding in specific cases, competent local government and unimpeded self-help movements are needed. Countries prepared to lay the structural and political foundations for such institutions should have priority when support is provided.

9. Under the conditions of the 1990s the success of rural development will largely depend not only on the creation of an enabling policy environment but also on the skilful linking of sectoral programmes and related (para)projects. Essential areas of cooperation in the struggle for greater and more effective orientation towards poverty are land and tenure reforms, rural financial systems, basic social services and social security systems. Experience has shown that, if rural growth processes are to benefit the mass of the population, solutions must also be found to major structural and social policy problems. Where small farmers enjoy permanently secure land use rights, they are also prepared to invest in production. Their actual investment behaviour largely depends on simple and unbureaucratic access to financial resources and on the possibility of keeping investment risks within bounds; self-help-oriented rural financial systems that also encourage saving could contribute to the elimination of existing constraints. Where small farmers - and the same applies to the landless - recognize business opportunities, they are willing to seize them. Practical primary education not only improves self-confidence but also helps to develop creative and

marketable skills. As many of the rural poor live in a very unhealthy environment, are constantly at risk within their own families and the village community and have to cope with emergency situations as they arise, considerable importance must be attached to adequate primary health care and stable social security. Unless there is an effective social safety net, the productive potential of the rural poor cannot be adequately tapped. Rural development projects and programmes have attached too little importance to these urgent structural and social policy tasks in the past. This must be rectified as a matter of urgency.

10. Development cooperation for rural areas must evolve from project-centred, isolated intervention to programme-centred interactive cooperation if the frustrations of the past are not to be repeated. Besides central administrations and local (political) government bodies, the (self-help) organizations of the rural population should be chosen as partners. Raising the level of intervention and linking programme promotion and individual projects will, however, also require some basic rethinking on the part of the donors, i.e. they must abandon practices of which they have grown fond, such as uncoordinated aid policies in which their own profile or supply interests are more important than the actual needs of the country concerned. In particular, fresh efforts must be made to ensure that proposed programmes and concepts and of project implementation strategies continue to be coordinated with partners. The policy dialogue should preferably be decentralized, i.e. the donors' local representatives should have more professional and decision-making competence so that they may react quickly, flexibly and comprehensively to political and administrative developments in partner countries. An on-going dialogue with social groups and associations is likely to be possible only through competent local representation that carries some weight.

## Introduction: International Criticism of Rural Development Projects

Integrated rural development is under discussion. The World Bank, whose former President McNamarra initiated rural development at international level as an approach to combating poverty with his famous speech in Nairobi in September 1973, is now its leading critic. A broadly based evaluation of 20 years of rural development as a focal area of activity is crushing in its condemnation of the majority of the World Bank's own projects.

Criticism of the various approaches to rural development is by no means new, however. The integrated approaches in particular have always been highly controversial. To put it rather simplistically, the two sides in the debate have been (agricultural) economists with neoclassical leanings and social scientists with their minds on the theory of systems. One of the critical areas of conflict in the implementation of rural development programmes was identified at an early stage: "A substantial allocation of essential resources to social services frequently occurs at the cost of more immediately productive investments in rural areas, and therefore, may prove self-defeating in the long run."[1] The complexity of integrated approaches and their manageability were also constantly on the agenda.[2]

The current debate on the high and low points of rural development must, however, be seen in the wider context of the permanent economic crises with which the poor countries, especially in Africa, are having to contend. Fundamental objections are raised to project-centred technical cooperation, whose limited problem-solving ability is now universally criticized. The drift of the criticism is that, although the many individual projects set isolated examples, encourage structural changes within a small area and have doubtless helped to improve the living conditions of a limited number of beneficiaries, unfavourable macrosocial, economic, fiscal and political structures have prevented them from having a wider impact and from being sustainable.

There are many project reports and evaluations to show that this is due not only to the now literally counterproductive environment but also to weakness in conception and in the organization of implementation.[3] The shortcomings that have been identified are generally grouped under the headings of lack of participation, non-sustainability, high unsecured fiscal follow-up costs, parallel organizational structures, inappropriate technology and limited impact on target groups.

The future of rural development as an instrument of development policy is likely to depend not only on an appropriate adjustment of the promotion concept at project level which helps to increase effectiveness and to improve the input-output ratio but also on action at sectoral and macro level specifically designed to benefit poor rural groups.

From past experience it can be assumed that adjustments at project level must be geared primarily to doing more to assist implementing agencies, to developing organizations and to promoting self-help, aspects which have dominated the development policy agenda for some considerable time, while structure-forming programmes are added at a higher level of intervention. It is particularly important in this context

- to strengthen the intermediate level between farmers and government,
- to decentralize the administration and to mobilize local resources,
- to develop financial services for groups who have traditionally been ineligible to borrow from banks and
- to provide social safeguards against losses of income and such risks as disease, invalidity, age and losses of harvest and livestock.

This study is intended as a contribution to the debate on the future of rural development and seeks to shed light on the area of conflict between a necessary modification of the project approach and sectoral programming for specific target groups.

The first part attempts a provisional appraisal based on a critical review of 40 years of rural development, while the second shows how action at the various levels must dovetail if future programmes for the promotion of rural development are to be more effective.

# I The High and Low Points of Integrated Rural Development (IRD)

## 1 Growing Poverty Despite Modernization and Economic Growth - Rural Development as a Strategy

### Development and growth through social mobilization

The history of integrated approaches to rural development begins in the 1950s with such social mobilization strategies as Community Development (CD). CD was paradigmatic of the then debate on development theory, which called for comprehensive social and economic modernization. It was argued that economic and social backwardness was largely caused by endogenous factors and was perpetuated as a vicious circle. Traditional societies and cultures met neither the psychological, social and political nor the economic requirements for progress. Besides social structures and patterns of behaviour that inhibited progress, it was such economic barriers as a shortage of capital, the absence of modern sectors (especially manufacturing industries) and the smallness of internal markets that caused and sustained backwardness. To break the vicious circles of underdevelopment and poverty, both investment in the development of modern industries (economic growth) and social change (acculturation) were needed.

CD therefore invariably focused on the village, with its traditional social order and patterns of behaviour. The measures taken related to education, the promotion of agricultural production, the improvement of infrastructure, and hygiene and health. The aim was to bring about comprehensive social change with a view to eliminating hunger, disease and ignorance. To initiate, control and oversee this social change, CD had the Village Level Worker, a village adviser. His task was to mobilize, explain, educate, advise and help.

From India, where CD remained - with various adjustments and additions - the dominant national rural development programme for five planning periods (1951 - 1976), it set off on a short triumphal march through the countries of Asia, Africa and Latin America. In various locally adapted forms, CD had spread to over 60 countries when it reached its peak in about 1960, by which time it was again being abandoned by the first countries to have adopted it. "Community Development had promised much, yet delivered little."[4]

As the agricultural reforms did nothing to change the traditional balance of power and distribution of land in the villages and the administrative structures did not allow of any serious moves towards grass-roots-oriented development through decentralization, the success of any socio-technological concept was bound to be no more than partial. Thus Village Level Workers were all too easily confronted with divergent group interests in the village and usually sided with those who had influence and property. Their inadequate professional and even human qualities in some cases and the absence of political and financial support from government also limited the effectiveness of these approaches.

Even in India, which, as the country where the movement had been born, long remained committed to the CD programme, its achievements were assessed in sobering terms:

> "Undeniably CD had created in its early years a great surge of hope and enthusiasm among the rank and file of the CD workers and villagers ... it must also be said that after its

first decade of operations the CD Programme had not significantly altered the basic conditions of rural life in India. Abject poverty, malnutrition and ill health and above all India's worsening food crisis had but one remedy - a larger production from the land."[5]

## Agricultural growth through the "Green Revolution"

In view of the mounting food crises, i.e. the widening gap between the growth of agricultural production and population growth, especially in Asia, supply-oriented promotional policies were more likely to meet the challenges of the Asian drama (Myrdal) than broadly based village development programmes. The ground for successful production orientation in agriculture had already been prepared in theory and practice in the mid-1960s:

- A reassessment of the role of the agricultural sector in the economic theory of development, inseparably associated with studies by Jorgensen, Ranis and Fei, Johnston and Mellor[6] and Schultz's hypothesis of the "poor but efficient farmer",[7] resulted in a greater commitment to the agricultural sector and so to sectoral agricultural promotional policies, with the emphasis on extension and research, the improvement of input supply, the development of agricultural credit systems and the creation of incentive systems.
- Breakthroughs were achieved in grain production with new high-yielding varieties, which formed the technological basis of the Green Revolution.[8]

In India, Pakistan, Indonesia and the Philippines, for example, the new agricultural development policy launched in the 1960s initiated a yield revolution on an unforeseen scale, helping countries to become self-sufficient and improving their food security. Outside Asia's irrigated agriculture the attainable advances in production initially kept pace with population growth. Minimum package programmes, designed to enable even small farms to increase their productivity by giving them access to cheap, i.e. subsidized, seed and fertilizers, were relatively successful in certain African countries, such as Ethiopia, but their effectiveness was limited by the unfavourable land tenure system (distribution of land ownership), institutional constraints (agricultural financing and banking systems; agricultural research) and the absence or weakness of production incentives (pricing policy).

The remarkable macroeconomic growth also achieved during this period painted far too rosy a picture of the actual situation since, despite all that has seemingly been achieved, it is becoming increasingly apparent that general modernization and growth policies produce the hoped-for welfare gains on only a limited scale and that these gains do not extend to poor population groups, especially those in rural areas.

## Disputed distribution effects of the "Green Revolution"

The undeniable gains in agricultural production in the Indian subcontinent due to the Green Revolution triggered a long debate on their social consequences. While the innovation package consisting of seed, fertilizer and technology appeared to meet all the requirements of a technology that was effective regardless of scale, it was primarily the large and medium-sized farms that benefited in the initial stages. Although it could also be applied to small farms, given the divisibility of the most important factors of production, they lagged far behind, with the result that the new technology further exacerbated the already marked social differentiation in rural areas. Does the new technology contribute to a differentiation of economic performance or does it intensify social polarization, it was pointedly asked. Both undoubtedly occurred in the process of technological change; it was already becoming clear at a relatively early stage that as a rule small farmers were put at a disadvantage primarily by unequal access to complementary factors of production (irrigation water, pumps, electricity) and services (agricultural credit) and by unfavourable land tenure systems (uncertain tenancy arrangements).[9] In India a Small

Farmers Development Agency (SFDA, in operation since 1969/70) was established in an attempt to enable small farms to derive appropriate benefit from the fruits of the new technology by promoting input supply, irrigation and marketing.[10]

Outside the debate on the effects of the Green Revolution it was realized, thanks partly to a wide range of empirical studies, that small and medium-sized family farms in particular have a potential that can be tapped and needs to be promoted. On the other hand, the really poor, those with no resources of their own, were largely overlooked and so de facto excluded.

## The birth of target-group-oriented rural programmes

McNamarra's famous address at the World Bank's annual conference in Nairobi in 1973 made the general public aware that 40 % of the world population are poor and live mostly in rural areas: rural development became a programme and strategy.

> "Rural development is a strategy designed to improve the economic and social life of a specific group of people - the rural poor. ... The objectives of rural development, therefore, extend beyond any particular sector. They encompass improved productivity, increased employment and thus higher incomes for target groups, as well as minimum acceptable levels of food, shelter, education and health."[11]

Despite being explicitly geared to alleviating poverty, the World Bank's rural development strategy was "from the beginning a smallholder project-based strategy (concentrating on those with productive assets) with only incidental benefits for the 'poorest of the poor' (laborers and the landless without productive assets)."[12] This was not surprising, however, for how was the World Bank to overcome the obvious inconsistency between poverty orientation and its mandate, as defined in its Articles of Agreement, to design projects that are economic and profitable? To reduce this conflict between conceptual aim and project reality, promotion was concentrated on the poor who had access to land; orientation towards this target group distinguished rural development projects from their predecessors, agricultural development projects. For monitoring purposes it was laid down that any agricultural project in which more than 50 % of the direct returns were intended to benefit poor target groups was to be managed as a rural development project. However, the importance of Area Development Projects in the rural development portfolio also grew. These projects have a regional bias, their aim being to increase the productivity of the bulk of small rural producers in provinces, districts or water catchment areas, where infrastructure and institutions are usually poorly developed. "They represent the heart of the rural development experience as originally proposed."[13]

The Bank, being a multinational institution, remained "neutral in structural policy terms" in the 1970s; although it wanted to increase productivity of small farms, it was averse to becoming involved in agricultural structural policy. It thus assisted programmes for small farmers in "both pre-revolutionary and post-revolutionary regimes - notably in Ethiopia, Nicaragua, Chile, Laos, and Congo."[14]

Besides the World Bank, it was above all the International Labour Organization (ILO) with its World Employment Programme that provided the empirical evidence in the form of broadly based country case studies to corroborate the doubts about the linear modernization approaches and made for greater awareness of the complexity of the situation. Attention increasingly focused on the phenomena of regional and seasonal underemployment and of the low productivity of labour due to the absence of complementary means of production, on the rudimentary factor and product markets and on the simple know-how transfer models of development cooperation, still usually known as development aid at this time.

## The rural development strategy takes shape

At the same time, another change of academic paradigm occurred, leading to a re-evaluation of the interaction between rural and urban sectors and a more discriminating analysis of agriculture's role in the processes of sectoral transformation in countries undergoing catch-up development. Transformation, assumed to be friction-free in the two-sector models of the 1960s, was abandoned in favour of the view that markets which function more or less well deplete factors and that the frame-work of political institutions (economic system) is decisive. The wide range of allocative efficiency that can be observed empirically is singled out as a central theme, and factor X, to which different institutional and management- and training-related effectiveness was allotted at that time, was introduced to explain highly divergent degrees of factor efficiency. Innovation research switched from models of the diffusion approach to aspects of endogenous innovation processes. The direction and momentum of agricultural change, for example, was attributed to advances induced institutionally in engineering, technology and organization: agricultural research and extension and the setting of relative agricultural prices played a central role in this context. Ways of achieving a scientifically oriented form of agriculture were indicated.[15]

This superstructure justified an agricultural and rural development strategy with the following main features:

- strategic entry points are the two relatively labour-intensive sectors agriculture and small-scale rural industry producing for local demand;

- promotion should focus on small farms rather than large ones because the abundance of small farms gives industrial and urban development a more profound impetus;

- for supply and linkage reasons priority is given to the promotion of food production;

- increased efforts to improve the substantive and organizational efficiency of agricultural research and extension;
- improvement of agricultural credit facilities, basic education (functional literacy) and the teaching of technical skills;
- the labour-intensive development and expansion of rural and agricultural infrastructure;
- such special programmes for poor groups as food-for-work, food stamps and fair price shops.

These strategic datum lines are to be found in both bilateral and multilateral projects that differ in sectoral and geographical scope and levels of intervention. The wide range of activities, whose planning and implementation approaches lag some three to five years behind the current state of the international debate, as reflected in the specialized journals, extend from those of the specialist adviser in the agriculture ministry through the organization of agricultural extension services to the rural composite, multisectoral project with a regional mandate.

## Competition among systems induces strategic adjustments

The greater interest taken in agriculture and the growing number of and financial weight carried by rural development projects explicitly intended to alleviate poverty formed part of a redistribution-with-growth strategy[16] whose academic and political protagonists cherished the hope that the inherent conflicts between the growth and distribution objectives in the development process could be sufficiently moderated for poverty to be limited and eventually overcome. Everyone was convinced that poverty itself must be tackled, since welfare gains achieved through growth were not trickling down as hoped. However - thus the lessons learnt from past growth processes - poverty must be so tackled that the macroeconomic growth process was not jeopardized, since without economic growth there could be no redistribution. Thus growth and equity was a reforming, not a radical approach; only a minority seriously considered a Marxist approach to be a viable alternative. Socialism as utopia seemed attractive to some, but it was no longer a desirable model because of the political distortions and repression in the majority of socialist developing countries, even though China initially appeared to have emerged as the victor from the real experiment of the competition between its and India's system in the mid-1970s. For China had shown that hunger and mass poverty could be controlled or were, indeed, things of the past. In contrast, growing underemployment, unemployment, poverty and hardship in developing countries organized along market economy lines, especially in South Asia and Africa, were leading to increasing migration from the land, with destabilizing consequences for rural and urban areas. It was only logical, then, that the new strategies should focus on rural areas and agriculture.

## Global crises, crisis management and the international economic order

At global level high population growth led to mounting concern about security of world food supplies. This resulted, among other things, in a world food conference (Rome, 1974) which agreed on a three-point strategy: an increase in production in the developing countries, an international early-warning and buffer-stock system as a hedge against unexpected losses of production, and improved distribution of food in the form of food aid. The newly created International Fund for Agricultural Development (IFAD) was to close the diagnosed investment gap of some US $ 1.5 to 2 billion p.a. in the developing countries' agricultural sectors. The newly established World Food Council was to coordinate international efforts to improve food security.

The explosive rise in oil prices after the Arab-Israeli Yom Kippur war, the subsequent dramatic deterioration of the growth prospects of many developing countries and the continuing decline of the international terms of trade led to increasing pressure from the developing countries for a new and more just international economic order; they thus raised the questions of growth and fair distribution to a global level. The links between national and international development and between the world market and domestic markets were more clearly recognized, and critical analyses were made of their implications for the project level. It was in this context that the basic needs strategy was formulated.

## 2 The Basic Needs Strategy and IRD - Grass-Roots-Oriented Development Despite an Unfavourable Environment

### Basic needs strategy

The call for a basic needs strategy[17] to overcome poverty that was voiced at the ILO's World Employment Conference in 1976 also signified the rejection of a distribution-oriented growth policy, which it was claimed was incapable on its own of satisfying the basic needs of those living in absolute poverty. What was needed, therefore, was a policy that provided sufficient essential goods and services, if necessary with direct government intervention where markets and commercial promotional policies failed.

The undisputed standard catalogue of the basic needs to be satisfied lists food, clean drinking water, health, clothing, housing and education; it is occasionally extended to include such non-material basic needs as self-determination, security and cultural identity. Although the sustained satisfaction of basic needs requires a reorientation of the whole of social and economic policy - which radical proponents of the basic needs strategy demanded from the outset - practical efforts under a basic needs policy focus primarily on agriculture and rural areas, both because that is where the vast majority of the absolute poor live and because the resource endowment of many poor countries (the exceptions being the oil- and mineral-rich) indicates that agriculture may make the largest productive contribution to meeting these needs.

The minimum requirements of a basic-needs-oriented development policy are considered to include the following:[18]

- development policy measures should be assessed for their direct contribution to the goal of satisfying basic needs (orientation towards needs);
- development measures must be explicitly aimed at the poor sections of the population (orientation towards target groups);
- measures to meet basic needs should be so designed that, wherever possible, the poor contribute their own labour and productive efforts (orientation towards production);
- the beneficiaries should be appropriately involved in planning and implementation so that they themselves determine their development (participation).

The objections to the basic needs strategy vary according to the academic or political viewpoint of the critics of this approach:

- it is a social assistance approach, which is not practicable since social policy cannot be financed indefinitely without productive growth;
- where the basic needs approach relates to rural areas, it is a new version of Community Development, which has been a total failure;
- the developing countries oppose this strategy because they see it as a ploy by the western donors to distract attention from legitimate demands for a New International Economic Order;
- the approach is bound to fail because the socio-economic structures and the political environment will not allow a basic-needs-oriented development policy to be widely implemented.

While the objections that the basic needs approach is primarily a social assistance concept or a new version of the unsuccessful CD, with references to the constituent elements of production orientation and target group orientation in the design of measures and programmes, are fairly easy to refute, the governments of the developing countries are indeed sceptical about this new strategy. Its proponents in donor and recipient countries must also accept that socio-economic structures and the political environment seriously restrict the scope for basic-needs-oriented programmes.

As the basic needs strategy reveals a conscious preference for agricultural and rural development, many developing countries with industrial ambitions and visions of development with an urban industrial bias tend to see it as an attempt by the old industrialized countries to refuse them the concessions in international trade and the transfer of know-how and information they need if they are to make industrial progress.

## Integrated rural development - difficulties with a concept

Under the influence of the international debate on the alleviation of poverty and the satisfaction of basic needs the rural development approaches (see Chapter 1) assumed practical form in the Integrated Rural Development (IRD) concept. While it is difficult to say precisely when and where IRD came into being, the concept matured at international level in the early 1970s.[19] Prior to this, rural development concepts had been discussed and used in the planning and implementation of development projects to justify action that can be defined as follows:

- "target-group-related" rural development: projects for small farmers (rationale: as a rule, only "progressive farmers" introduce general agricultural innovations, most small farmers being inhibited by factors peculiar to them as a group);
- "poverty-oriented" rural development: projects for the rural poor (rationale: the macro strategy of "redistribution with growth"

needs to be complemented programmes specifically designed to benefit poor rural groups, who usually remain outside the "mainstream economy";

- "basic needs-oriented" rural development: the productive mobilization of the particularly needy sections of the population and the provision of such public services as health, hygiene, drinking water, education, transport and cultural institutions (rationale: reducing the marginality of the poor through integration is possible only if development policy is consciously geared to the priority of satisfying basic needs).

Very early and influential examples of the new type of integrated rural development approaches, such as the "Comilla approach"[20] and the "CADU project" (Chilalo Agricultural Development Unit, a project of the Swedish SIDA in Ethiopia) gave planners and administrators of rural development projects a great deal of encouragement as they undertook the difficult process of formulating clearer objectives.

In the early stages of its development IRD was characterized by a lack of conceptual clarity and of operationality. "The diversity and the vagueness of definitions of IRD is not merely a semantic issue. Rather it reflects the lack of an understanding of rural problems and a failure to agree upon means as well as ends in ameliorating them."[21] It is therefore hardly surprising that - being committed to a noble objective - it was quickly reduced to a slogan or wrongly taken to be a panacea.

The IRD approach does not claim to be based in development theory, but rather describes a type of action whose constituent ingredients are:

- a multisectoral approach (productive and social sectors);
- orientation towards poor rural target groups (e.g. the landless, farm workers, marginalized farmers, women, the handicapped);

- social mobilization and active involvement of the beneficiaries (participation);
- (a regional bias, e.g. to a district, province or watershed).

While Rural Development consisted of almost any combination of measures to increase agricultural productivity and production and of complementary infrastructure measures and social action, Integrated Rural Development has always suffered from the absence of conclusive answers to the question: what and who should be integrated into rural development programmes and/or projects, for what purpose and how? The interpretations cover a wide range, from a social-normative view - (re)integration of marginalized rural groups and individuals into the process of developing society as a whole - to a pragmatic, management-related view,[22] meaning the combination of various components to give a coordinated multisectoral programme package under uniform management.

## What does "integrated" mean in integrated rural development?

In the paper he delivered to the International Conference of Agricultural Economists held in Nairobi in 1976 M. Yudelman,[23] for many years director of the World Bank's Agriculture and Rural Development Department, presented three pragmatic, management-related views on the question of integrated, which are reflected in specific types of project undertaken by the World Bank:

- integration means the coordinated provision of the inputs and complementary services that small farmers need to increase production (package programmes or minimum package programmes);
- measures to promote agricultural production are integrated into socio-economically oriented projects (e.g. Community Development, rural public works programmes, basic education and skill development);

- agricultural and non-agricultural activities are combined in a comprehensive development programme (comprehensive approach: a) Coordinated National Programmes; b) Area Development Schemes).

Besides ensuring the linking and coordination of instruments that is desirable and can be deduced from the logic of farm production methods and so making action both more efficient and more effective, integrated approaches are deemed to have synergy effects because the whole happens to be greater than the sum of its parts. The possibility of high organizational costs sapping the synergy effect was, however, recognized and enunciated at an early stage. As integrated, i.e. multisectoral, projects and programmes have to contend with sectorally and functionally divided administrations (ministries, departments, etc.), there is a need to create horizontally linked - integrated? - administrative structures, which may clash with the vertical chains of administration that have evolved over the years.[24]

The social-normative view that the goal of rural development is the social and economic integration of socially and economically marginalized groups is gaining in importance as the overriding objective or model, without making any direct operational claims. Particularly controversial is the question of the political structures needed for the effective implementation of poverty-oriented approaches. Although the need for structural reforms (e.g. of land tenure systems and of market and pricing policies) is endorsed in principle, a radical position which insists that basic-needs-oriented or poverty-oriented measures should not be taken until structures have been drastically changed is rejected in favour of a pragmatic position: "... taking full advantage of and increasing the scope provided by development policy for development at local and project level is seen as the pragmatic starting-point for 'rural development'."[25]

In defining the principles of IRD,[26] the German Federal Ministry for Economic Cooperation[27] states that rural development projects are characterized by the following elements:

- orientation towards target groups,
- the principle of participation,
- a multisectoral approach,
- gradual progress in planning, preparation and implementation and
- the satisfaction of basic needs.

The goal of rural development is

> "a lasting effective improvement in the living conditions of people - men, women and children - in rural regions on the basis of
>
> - economic and social self-determination, with account taken of cultural independence,
> - agriculture, forestry and fisheries that are diversified and appropriate to the location and do not deplete resources,
> - a multisectoral approach,
> - efficient physical and social infrastructure and
> - decentralized craft and small-scale industrial production firms."[28]

The genesis of the IRD approach shows that the basic needs strategy and rural development are inseparably linked, that they form an end-means relationship, but that they are encumbered with the odium of lacking a basis in theory, i.e. they do not offer a concise approach to explaining the phenomenon they seek to eliminate: mass poverty in the developing countries. Hence the relative popularity of the interpretation, hence too the often fundamental criticism voiced by proponents of an autocentric development approach based on the dependence theory.[29] The question that arises for them is whether integrated rural development has any chance at all of succeeding while the traditional power, economic and cultural structures remain in place. As the link between rural development and what happens in the political and economic macro structure also remains unclear in the

IRD concept, it is virtually impossible to deduce a consistent pattern of action.

The agricultural economists' criticism is voiced with particular clarity by Ruthenberg:[30]

> "Efforts in rural development at village level are hopeless propositions unless they stimulate cooperation and support from that third of the farmers that commands about two-thirds of the land. The idea found in a great number of publications on rural development of organising the rural poor (small farmers and landless) without or even against the 'rich' farmers, is realistic only where collective farming is the objective, and such a move would probably meet the opposition of the majority of all farmers."

He also describes the approach in itself as unrealistic because it is too complex in design and seeks to achieve too many, even conflicting objectives within the organization of one project.

Academic circles reacted to the early criticism both by attempting to give the concept a comprehensive theoretical base, with systems analysis recognized as a useful instrument,[31] and by operationalizing IRD as a methodological approach[32] in order to identify time- and space-related and functionally oriented investment patterns for all decision-making levels.[33] Development practitioners also played an active part in the attempts at operationalization.

## 3 Regional Rural Development (RRD) as a Prototype of Rural Development Projects

The new-style Area Development Schemes (World Bank and USAID) and the Regional Rural Development (RRD) projects are considered to be prototypes of the operationalization of the IRD philosophy. Its pragmatic operationalization is based on three fundamental conclusions:

a) Given the complex links between the sustainable and wide-spread satisfaction of basic needs, target group orientation and participation, a carefully coordinated case-by-case approach is needed. A number of simple principles can, however, be established:
   - satisfaction of basic needs: the many and varied structural causes of inadequate provision call for a multi-sectoral approach to problem-solving;
   - target group orientation: as the planned packages of measures must have an impact on the poor, the latter must not be too heterogeneous;
   - participation: if those directly concerned are to be actively involved in the planning and implementation of measures, there must be close cooperation with their representatives.

b) Where practical implementation is concerned, it follows from this that, if they are to make a significant contribution to the alleviation of poverty, projects should be implemented primarily in small geographical regions that go some way towards meeting the criteria of social and ethnic homogeneity and are regarded as poor regions in the national context.

c) The conditions required if the implementation of a project agreed by states is to succeed are most likely to exist in an administrative unit such as a district, region or province, provided that at least some scope is allowed for the decentralized planning and implementation of development projects.

## The World Bank's and USAID's regional development projects

In the World Bank's rural development programme a distinction can be made between two categories of rural development project: on the one hand, modified agricultural projects and, on the other, the new-style projects, the most important elements of which can be described as follows:

> 1. "They are designed to benefit large numbers of rural poor, while earning an economic rate of return, that is, at least equal to the opportunity cost of capital.
>
> 2. They are comprehensive in their approach to small-scale agriculture and provide a balance between directly productive and other components (where inclusion of the latter is appropriate).
>
> 3. They have a low enough cost per beneficiary, so that they could be extended to other areas, given the availability of additional resources."[34]

Of the 529 rural development projects financed by the World Bank from 1974 to 1986, 40 % were classified by the Bank itself as Area Development Projects, about 27 % as irrigation projects and 15 % as credit projects.[35] It is the new-style Area Development Projects that have become the World Bank's leading type of integrated rural development (in various cases irrigation projects might also be reclassified as Area Development Projects).

Much the same can be said of USAID, where it is above all rural development projects with an explicit regional bias that most clearly represent the new type of multisectoral, integrated project. USAID defines its IRD projects as having (a) a limited, clearly defined geographical mandate, (b) a multi-sectoral mandate, (c) a coordinated approach to the provision of goods and services for the local population and (d) a certain degree of participation by the beneficiaries.[36]

## From regional agricultural projects to regional rural development

In the mid-1960s there emerged within the framework of German bilateral technical cooperation the type of regional agricultural project that was to develop into the integrated regional rural development project over a period of ten years. Three of the most important cases, whose project history illustrates the conceptualization of the German version of integrated rural development, were the regional multisectoral project in Paktia/Afghanistan, the regional project in Salima/Malawi and the regional agricultural project in West Sumatra/Indonesia.

Planning for the Paktia and Salima projects began in 1963 and 1965 respectively. The Province of Paktia is a backward, politically difficult mountainous region half way along Afghanistan's eastern frontier. From 1966 the Afghan Paktia Development Authority was assisted with advice and material support. The German contribution consisted of several, formally separate projects: agriculture, forestry, infrastructure and promotion of the crafts. The coordination of the various projects was the responsibility of the regional development authority. In a brief description of the agricultural project the planning principles were defined as follows: "Capital-extensive and labour-intensive project concept, activation of local initiative and self-help, regional development on the broadest possible basis to encourage competition, while preventing tribal rivalries".[37]

While the Paktia project was a composite, multisectoral regional project, the Salima project was a "regional project based on the principle of simultaneously taking all the measures needed to increase agricultural production as a package in a defined area."[38] The project was later renamed the Central Region Lakeshore Development Project to show that it was a multisectoral regional development project. On the basis of the experience previously gained in Paktia and Salima, preparations for the West Sumatra Agricultural Development Project began in 1969. It was conceived as a regional planning and implementation project, i.e. the establishment of a regional plan formed part of the project mandate. "The intention is to promote the largely agricultural economy of the Province of West Sumatra with planning, selective development programmes and the opening up of new markets."[39]

Although each project was considered unique because of the specific local conditions and the planning philosophy of its sponsors, each was also influenced by similar projects in the same country or nearby: Paktia by such projects as Helman Valley/ Afghanistan (USAID) and Mandi/India (GAWI), and Salima/Malawi (GAWI/KfW) also by the World Bank's Lilongwe Land Development Project (from 1967). Perhaps the most influential projects in the international history of integrated rural development were Commilla/East Pakistan (Bangladesh), CADU/ Ethiopia, Puebla/Mexico and Lilongwe/ Malawi.

## Regional rural development and regional economic approaches

The genesis of integrated regional rural development projects was largely inductive (related to experience), although the actual plans were based on different approaches in agricultural and regional economic theory, imparted by the development planners involved. Almost all practical attempts to step up or launch socio-economic development in a given area - district, province, valley, etc. - with action from outside have begun with measures that directly and indirectly promote production on small farms.

The heterogeneity of natural resource endowment, the fact that in the early stages of agricultural development primary agricultural production very much depended on the location, and other regionally specific factors have meant that the pace of socio-economic development has been far from the same everywhere. Both the given spatial differentiation and the existing regional disparities thus confront the development planner with complex decision-making problems; for solving them, he needs to make not only functional or sectoral and time-related but also spatial calculations. While plans for programmes to promote agriculture (e.g. minimum package programmes) can be guided by models of certain farming systems appropriate to the existing physical and ecological factor endowment, comprehensive investment plans require an evaluation both of the specific location and of monetary inputs and outputs.

In this context the derivation of the spatial decision-making calculation in planning is based on regional economic approaches.[40] The most influential approaches have so far been the theory of the growth poles, the theory of central places, the export-base model and the concept of agropolitan development. The regional rural development project type owes to these regional economic models not only a functional but also a territorial view of development, which formulates as its strategic approach to overcoming or at least alleviating regional disparities (rich vs poor regions) more inwardly oriented development (agropolitan development) rather than a hierarchical order in the area concerned (central places) and the economic integration of even remote regions into the worldwide division of labour (export-base model). However, the ideas underlying agropolitan development have yet to have any demonstrable influence on the first-generation projects (1965-1975).

## RRD projects as "flagships" of development cooperation in rural areas

Although regional rural development or area development can be regarded as the most mature implementation of the ideas behind integrated rural development, this type of project has never been able to attract more than a good third of all the development cooperation funds earmarked by major donors for the rural sectors. The rest has been spent on the planning and implementation of agricultural sectoral programmes and more specific subsectoral agricultural and rural projects.

Leaving aside the development policy concepts of the recipient countries, which have differed widely in their reactions to international offers to implement integrated rural development programmes, pragmatic decisions on the relative advantages of alternative project approaches have usually been based on the donors' analyses of constraints. Where the constraints are identified as existing primarily in such key services as credit supply and agricultural extension services, functionally oriented projects are appropriate. If, on the other hand, the weaknesses and obstacles in the agricultural system lie mainly in input supply (fertilizers, pesticides, tools, simple agricultural machinery), marketing and/or processing, sectoral projects are needed to eliminate the deficiencies. However, where infrastructure and public services are generally inadequate, an integrated rural approach with a regional bias is suitable.

For many poor countries these selection rules are unlikely to be adequate because of pronounced institutional weaknesses and a wide variety of constraints. Typically, a large number of agricultural and rural projects both of a functional and sectoral and of a regional nature are usually to be found in such countries. How far they together contribute to convincing problem-solving depends not only on the recipient countries' institutional absorptive and implementing capacity but to a great extent on effective donor coordination.

## Administrative constraints and project policy

By the late 1970s, with all the influential donors becoming increasingly committed to regional rural development, the growing number of integrated projects were already stretching the administrative and institutional capacities of many countries to their limits, well-known examples being Nigeria, Niger, Malawi, Kenya and Lesotho in Africa, Sri Lanka and the north-east of Brazil. In the World Bank's case at least, area development projects in African countries consequently began to give way to sectoral promotional programmes, especially in the areas of agricultural research and extension, in the early 1980s. Although such other donor organizations as USAID and Britain's ODA continued to back IRD, the number of new projects declined in the 1980s. The GTZ, on the other hand, saw regional rural development as a particularly important and effective means of honouring the Federal Government's political commitment to poverty alleviation: rules on the planning and implementation of poverty-oriented rural development projects appeared in 1978 and were followed in 1983 by guidelines entitled "Regional Rural Development", which contained a fairly binding definition of this type of project.

Nonetheless, it seems that, as the focus of efforts to achieve integrated rural development, regional rural development was destined to flourish internationally for only a brief period, since it had already passed its peak as a project type before maturing fully as a concept; major donors, headed by the World Bank, are turning to other project types, seeing functional and sectoral agricultural and rural projects as more promising and cost-effective. This is particularly true of the African countries, where the profound economic, social and political structural crisis is prompting a fundamental change in levels of action to macroeconomic and macropolitical adjustment programmes.

## 4 IRD/RRD - a Failed Approach?

Has integrated rural development and, with it, regional rural development failed as an approach or been, at best, an episode in the constantly changing pattern of development theory paradigms and development concepts? Since the mid-1980s, if not earlier, IRD has been increasingly regarded as a failure and attracted the same criticism as Community Development before it: it promises much, but delivers little. Besides radical proponents of a new agricultural growth strategy, it is a group of reform-minded critics who, while acknowledging the need for direct action to benefit the rural poor, disapprove of integrated approaches because their structural effectiveness has been extremely disappointing.

The most important arguments advanced in the international debate on IRD/RRD can be essentially narrowed down to the following two:

a) Multisectoral projects are too complex, too large and too cumbersome, and they are inappropriate to a sectorally structured administration; they are difficult to control and require the installation of project structures parallel to the sectoral administrations. This not only has an adverse impact on the effectiveness of projects but above all prevents institutional sustainability, since the administration in charge of the planning and implementation of development programmes, which should be empowered to implement projects on its own responsibility, is forced to play a minor role. Moreover, the project structures that emerge are usually top-heavy and absorb excessive manpower and financial resources for costly planning and coordination processes, to the detriment of the programmes.

b) As evaluations show, integrated projects have so far neither had a wide impact nor been sustainable, i.e. they have achieved neither the declared aim of effectively alleviating or eliminating rural poverty and impoverishment processes nor the sustained mobilization of human, organizational and natural resources. Most projects have in fact suffered because the improvement in the capacity for self-help has been very slight, the barriers to access to necessary services (markets, extension, credit, means of production) have been overcome only in isolated instances and the promotion and ecological stabilization of production systems has been inadequate. In short, they are far too technocratic, they are planned from the top down, and they leave no scope for organization by the people themselves.

The critics cite a number of difficulties as causes of the relative failure of integrated approaches. Besides conceptual weaknesses and shortcomings in implementation, they refer primarily to unfavourable social and political conditions in the project environment, resulting in frictional losses and reduced effectiveness. For analytical reasons it would perhaps be useful to begin by considering inherent weaknesses in the approaches and their implementation before those aspects of the political environment that threaten success and may even be counterproductive and their relationship with IRD/RRD projects are discussed.

## Conceptual weaknesses and implementation problems

Since they pursue a wide range of goals, the constant danger with IRD/RRD projects is that the formulation of objectives, the assignment of objectives and implementing methods and the sequencing of targets, methods and progress reviews will be relatively vague, unrealistic and not entirely consistent, the complexity of the management task thus tending to overtax the institutions, organizations and actors to whom it is entrusted. If projects are, moreover, implemented in regions which have limited natural resources and are afflicted by crisis and where rural poverty is therefore concentrated, difficulties and failure seem almost predestined.

However, past experience shows there to be many stages between manageable difficulties and total failure. The key elements are project organization, coordination and participation.

## Project organization and multisectoral coordination

If multiple objectives are to be achieved, coordinated action (policy cohesion) across sectoral boundaries is needed. Being inadequately coordinated horizontally, sectorally structured and usually inefficient to boot, the administrations of the developing countries are incapable of meeting this need. The relevant administrations are, moreover, centralistically structured and delegate little authority to their subordinate regional offices; this applies to the inadequate transfer of both planning and financial authority to lower regional levels. To overcome this failing, independent project organizations able to act with relative autonomy were usually established in the early days of IRD/RRD. As these autonomous project organizations gave rise to the familiar handing-over problems, alternative implementing structures were sought. Some IRD/RRD projects were attached to a lead ministry - often the agriculture ministry - while in other cases territorial authorities (e.g. provincial governments and district administrations or councils) and rural development agencies or ministries took responsibility for their implementation. Probably the most common form of organization was the attachment of projects to a lead ministry. How the chosen implementing structure performed depended in each case on its planning, implementation, coordination and budgetary competence.

To generalize, it can be said that, even where an implementing agency has considerable formal authority in a centralistically administered country, the vertical structures controlled from headquarters virtually rule out the coordination and cooperation at regional level which a project needs and from which it benefits. In this situation a territorial authority acting as an implementing organization will be tilting at windmills in its dealings with the normally arrogant line agencies. The required sustainability of project impacts is unlikely to be achieved. Where structures are decentralized and ensure the delegation of powers to regional levels, the attachment of IRD/RRD projects to a territorial authority should augur well for their success. Such an authority is in the best position to ensure that planning and implementation are carefully attuned to the target group and that there can be timely and appropriate coordination.

## Participation and self-organization

As long as projects are regarded as means of taking action that make the beneficiaries the passive objects of decision-making, involved in the creation of planned outputs as quantity adjusters, self-sustaining economic and socio-organizational processes are hardly likely to occur. This conception of projects, which is all too common, despite rhetoric to the contrary, also conforms to a school of thought on modernization that reduces the phenomenon of backwardness to a problem of technology transfer. Many examples can thus be given to show that, while the transfer of inappropriate imported technical, organizational and institutional solutions may ease the real pressure of problems in IRD/RRD projects - as in sectoral projects - the desirable sustained assimilation by those directly concerned does not occur. Administered innovations,[41] i.e. innovations decreed from above, have, as many projects have shown, limited viability. If IRD/RRD projects are to act as effective promoters of innovations that improve performance, i.e. as innovation agencies,[42] which is the declared aim of technical cooperation, the provision of appropriate technical, organizational and institutional problem-solving machinery is a priority task for projects and is possible only if the beneficiaries and their organizations play an active part in the decision-making processes. A project's main role is then to act as a facilitator or catalyst that seeks to set processes in motion. This is the case with such strategic approaches as institution- and capacity-building and participatory action research. Their aim is to develop, together with the beneficiaries, suitable solutions to problems, which will become more widely and sustainably established as a result of socially appropriate transfer systems than administered innovations.

## Planning styles - blueprints versus learning processes

Despite all the complexity of the tasks to be performed by IRD/RRD projects, planning techniques and styles which pre-supposed that both the ends and the means needed to achieve them were fully or at least adequately known and manageable remained in fashion for a long time. It was a relatively simple business to convert the blueprints that were drawn up for projects with good intentions, but without the participation of the beneficiaries or those directly concerned, into practicable, transparent operational plans, and they also made it easy for both the sponsors and the evaluators appointed to monitor progress. Designing projects as learning processes and gearing plans to this objective was far from successful even after blueprint planning had failed.

Although the ZOPP (objective-oriented project planning) method developed in German development cooperation enables a structured dialogue on problems, goals and possible solutions to be conducted with the various parties concerned, it translates this dialogue into a vertical causal form of logic, which can only be as good as the understanding of problems and theory of those participating in the ZOPP workshop. Process-oriented and interactive learning are reduced to ends-and-means relationships or if-then sentences. While local adaptations of the ZOPP method[43] prove that this planning procedure can certainly be geared to specific target groups and used participatorily, the planning process should be made generally more open so that it may react more appropriately to the views and strengths of those concerned and sustain the momentum of change once it has begun.

A major step towards more open planning was the introduction of two- to three-year planning and orientation phases into the cycle of German RRD projects. They provide opportunities for both an in-depth analysis of the problems and a tentative search for suitable problem-solving approaches.

## Conflicts in implementation - project personnel under pressure to succeed

The personnel employed undoubtedly represent a serious weakness in the implementation of IRD/RRD projects. Despite their unquestioned professional qualifications, they frequently lack the ability and perhaps the willingness to conduct an on-going dialogue with the various parties involved (those concerned and the potential beneficiaries) on the direction of the change that is needed for the adjustment of objectives, resources and methods in the form of an open planning process. The real goal of technical cooperation, an improvement in the competence of the target group to take action through the removal of any obstacles that stand in its way, i.e. helping people to help themselves, is unlikely to be achieved by direct action: it calls for the self-organization of those concerned, which must evolve since it can hardly be prescribed. Time being a scarce commodity, project personnel come under pressure to succeed, with sponsors and evaluators usually expecting evidence of quantifiable, measurable and visible progress that meets their own (western) standards. Both project personnel and sponsors and evaluators should be made even more aware of these links if the sustainability of the effects of projects is to be regarded as the real and true yard-stick of success.

## IRD/RRD in a difficult economic, institutional and political environment

Besides the shortcomings in the conception and implementation of integrated approaches, unfavourable environmental aspects are singled out to explain the limited success of projects or their failure. Among the most commonly mentioned are:

### Authoritarian structures

- low public confidence in the legal system and limited civil freedoms;
- administrative and legal obstacles to the self-organization of target groups;
- social policy programmes that do not as a rule take sufficient account of poor population groups, with the result that they have little or no access to health and education services;
- structurally hampered access for poor population groups to inputs (especially land, agricultural inputs, financial services) and markets;

- the failure to regard women as the relevant producers they often are.

## Lack of performance incentives

- etatist regulatory and structural policies that are to the disadvantage of rural producers and the informal sector;
- low agricultural producer prices;
- poor absorptive capacity of local markets.

## Centralist administration

- administrative machinery geared primarily to fiscal and regulatory goals;
- absence or incompetence of local government;
- institutional inefficiency of government administration, with extensive corruption;
- poorly motivated public servants.

It is quite true that in the majority of poor countries development co-operation is confronted with a social and political environment which is hardly conducive to the achievement of the main aims of IRD/RRD: lasting and self-sustaining social, economic and ecological development. An added difficulty is that, almost by definition, IRD/RRD projects are implemented in disadvantaged, structurally weak regions, which are underdeveloped even and especially in the national context. Many are also security-sensitive border regions, where IRD/RRD projects are seen as counter-insurgency programmes (there being numerous examples in Asia and Africa).[44]

## Critical interfaces between projects and the policy environment

To determine how far specific project failures can be ascribed to an unfavourable environment, i.e. adverse economic and political structures, rather than home-made short-comings in planning and implementation and the long cherished illusion that the required structural reforms could be initiated and implemented from below by action at project level, they would have to be carefully analysed case by case.

The particularly critical interfaces between a project and the policy environment are:

choice of implementing agencies: if projects are the responsibility of inadequately decentralized government line agencies and their local representative bodies, the mere increase in their staff, financial and technical capacities to which projects typically give rise helps to create a situation in which policies can be even better imposed from above; this is true of both favourable and unfavourable cases. Little or no advantage can then be taken of the scope for creativity that otherwise exists;

externalization of planning: the selection and employment of promotional instruments is normally based on external analyses of constraints and is rarely the outcome of a joint process of learning by project personnel and those directly concerned. Conflicts of interest and incompatibilities only become visible later, usually as failures, which can be mitigated in an open planning process;

projects as enclaves: given their financial strength, IRD/RRD projects may neutralize an unfavourable environment for a time through subsidization and so ensure their own effectiveness at the cost of sustained funding (i.e. affordable follow-up costs).[45] This leads to the temporary creation of prosperous enclaves, whose lifespan is largely determined by the project cycle;

scope of the project approach: if the adjustment of macroeconomic aggregates, the structure of internal economic incentives and the social sectors constantly and systematically puts the rural and informal sectors at a disadvantage, projects can at best alleviate the adverse effects of the failure to mobilize resources, any misdirection of factors of production and growing poverty, but - despite subsidization - it is asking too much of them to make a lasting contribution to self-sustained development. The project approach (micro level) is then bound to fail, and the logical consequence would be to raise the level of action at least to the meso level.

## 5 Lessons Learnt from 40 Years of Rural Development - a Provisional Appraisal

After four decades of efforts to achieve rural development with various promotional approaches, one rapidly succeeding another, among them Community Development, the Green Revolution, integrated rural development and helping people to help themselves, with the level of action constantly changing - "self-help organizations at the level of villages, sectors and poor regions" - and accompanied by such more or less succinct strategy recommendations as modernization, expansion of research and extension services, programmes for specific target groups and organizational development, it is worth redefining the position and taking provisional stock (see Figure 1):

- What relatively sound conclusions can be drawn from the past 40 years of international efforts to achieve rural development and a lasting reduction in rural poverty?

- What direction should future thinking and action take to bring the goal of an appreciable reduction in mass rural poverty closer to achievement?

With all the necessary caution that generalizations require, the following findings and conclusions can be regarded as fairly sound and used for an adjustment of development cooperation and its rural development instruments:

a) Economic growth, and especially an increase in the productivity of the rural sectors, is essential if rural poverty is to be reduced. Although Green Revolution strategies, where they had an impact, as they did in South and South-East Asia, tended to make the rich richer, they did not make the poor poorer. In South Asia today local food production feeds twice as many people as 20 years ago. It is fundamentally important, however, for the quality of growth to be such that ecologically sustainable development is ensured.

b) Appropriate agricultural technologies geared specifically to the poor resource endowment and the social basis of small farming systems are still uncommon. It is only comparatively recently that the international agricultural research community has begun to consider important and typical small farming systems in its research on breeding and to learn to understand the socio-cultural peculiarities of small-scale farm production methods and the effects they have on small farmers' attitudes towards innovations.

c) Agricultural and social services are still barely within the reach of the rural poor despite numerous attempts to widen and improve the range. High priority should be given to action aimed specifically at the poor in this respect. Health care, adequate primary education and assured access to means of production and information are essential if economic activities are to succeed.

d) Agricultural reforms have failed to progress beyond the initial stages almost everywhere. While land reforms certainly cannot solve all poverty-related problems, easier access to land use rights and greater security for tenants can make significant contributions to the reduction of rural poverty.

e) Agricultural market and pricing policies were used for far too long to finance urban and industrial development by keeping producer prices low, pursuing inappropriate exchange rate policies and levying taxes on agricultural exports. In many African countries in particular the result was a decline in per capita agricultural production. Creating incentives by raising producer prices and liberalizing markets was therefore a policy adjustment that obviously needed to be made. As the price elasticity of supply from small farmers is very limited in the short to medium term because of the prevailing structural constraints affecting agricultural inputs, and yet as consumers they feel the full effect of price increases - less so, no doubt, than the urban poor, who are totally dependent on the market - this instrument should be used only in small doses. Gradually bringing about undistorted producer prices is undoubtedly in the interests of poor rural producers since they have the effect of increasing productivity.

f) Non-agricultural employment, or the linking of multiple income strategies, increasingly determines the attainable level of income. This is true not only of the landless but also of a rapidly growing number of small farmers. With population pressure high and ecological damage rising sharply and spreading to ever larger areas of agricultural and forestry land, non-agricultural employment and income-generating opportunities must be increased as a matter of urgency. This can be achieved only if people are better able to take the initiative and help themselves (confidence-building) and financial and economic policies are changed to permit the emergence of financial systems that benefit specific target groups and the gearing of promotion to the informal sectors.

g) An early reduction in population growth to a level at which economic and social progress is able to have a wide impact can be achieved only if there is a lasting reduction in poverty, and hardship and social security is guaranteed. High birth rates are primarily due to poverty, poor education, deficient health care and the absence of social security. Appropriate and effective protection against unavoidable risks to life and production, combined with improved access to social services, will not only

**Figure 1 - Main concerns of agricultural/rural development since the 1950s in the context of development theory and development policy**

| Main themes of the international debate | **Modernization of traditional agriculture** | **Growth of agricultural production** | **Alleviation of rural poverty** | **Strengthening the capacity for self-help** |
|---|---|---|---|---|
| Period | 1950s - 1960s | 1960s - 1970s | 1970s - 1980s | 1980s - early 1990s |
| Prevailing approaches to the promotion of agriculture/rural areas | **Community Development (CD)** | **Seed, fertilizer, technology (Green Revolution)** | **Integrated rural development (IRD/RRD)** | **Development of organizations (capacity-building)** |
| Typical level of action, beneficiaries | Village; rural community | Sectoral policies; agricultural service institutions | Poor regions (districts/ provinces), target groups | Self-help organizations |
| Recommended strategy | Overcome economic backwardness and traditional attitudes through modernization; IL model | Biological and technological progress can raise the level of productivity; this presupposes efficient agricultural extension services, agricultural research and agricultural credit | Neither modernization nor growth strategies can successfully alleviate poverty; therefore programmes specifically for small farmers and other poor rural groups | Sustainability of action is ensured only if beneficiaries are actively involved in the solution of their own problems |
| Historical and development policy context | Decolonization<br><br>successful growth processes in the industrialized countries;<br><br>development optimism;<br><br>belief that development can be planned;<br><br>Cold War: rivalry between market and planned economies | Regional food crises;<br><br>high population growth;<br><br>World Food Conference 1974;<br><br>Sahel disaster;<br><br>Decade of Women 1975 - 1985 | 700m rural poor and rising;<br><br>growing unemployment;<br><br>1st oil crisis 1973;<br><br>World Conference on Agricultural Reform and Rural Development (WCARRD) 1979;<br><br>2nd oil shock 1979/80 | World economic recession;<br><br>mounting debt crisis;<br><br>Africa's structural crisis;<br><br>exacerbation of environmental problems;<br><br>regional conflicts;<br><br>civil wars;<br><br>end of the Cold War |

Figure 1 (cont.) - **Main concerns of agricultural/rural development since the 1950s in the context of development theory and development policy**

| Leitmotif of development policy | **Catch-up development (industrialization)** | **Integration into the world market** | **Inward orientation** | **Structural adjustment** |
|---|---|---|---|---|
| General debate on development theory | Modernization theories:<br>- economic growth theories and "trickle down"<br>- theories of social change<br>- diffusion theory of innovation | Terms of trade debate;<br><br>dependence theories;<br><br>dissociation concepts | Redistribution with growth; ILO's employment-oriented concepts; satisfaction of basic needs; poverty-oriented strategies | Fundamental criticism of development policy; ecology and development; World Bank's neoliberal concepts ("get prices right") |
| Debate on the theory of agricultural development | **Industry as "leading sector"**<br><br>Pessimism about agricultural productivity; transfer of labour and capital from agriculture to modern sectors (double squeeze on agriculture) | **Traditional agriculture: "poor, but efficient"**<br><br>Major productivity gains possible in agriculture (high pay-off input model); theory-induced innovation (role of research and extension) | **Unimodal development (progressive modernization)**<br><br>Interaction: agriculture - industry; small farmer strategy; employment-oriented agricultural development | **Role of agricultural market and pricing policy**<br><br>Land tenure system and rural poverty; development of agricultural technology (farming systems research; organic farming) |

help to reduce the birth rate in the longer term, but even in the medium term result in the beneficiaries exhibiting additional creative and productive forces.

h) Democratization, i.e. the widest possible participation of the people in the formulation of political objectives and in the dialogue on development, is one of the keys to successful development projects and programmes. Only if the beneficiaries are actively involved in solving their own problems are measures likely to be sustainable.

i) The decentralization of the machinery of government and the strengthening of local self-administration are necessary if planning and action are to be needs-oriented and participatory. They are, moreover, helpful means of reducing the usual opposition to local taxes and levies, since local decision-making on and monitoring of the use of resources are then more easily ensured. Furthermore, decentralization makes for greater transparency of administrative activities. In practical terms decentralization means the development and expansion of local government structures. As it results in losses of power and control, unlike the transfer of administration downwards (deconcentration), there is usually considerable active and passive resistance from central institutions.

j) Development of intermediary organizations between government and people. The dialogue between government and people on the direction that socio-economic change should take and how it is to be achieved cannot be institutionalized easily without the involvement of such intermediary organizations as cooperatives and professional and other associations whose members are both individuals and small self-help groups close to the grass-roots level. Besides acting as a lobby, this third sector may perform certain public tasks and so ease the burden on government.

k) The geographical and political range and model nature of rural development projects is normally very limited. As a rule different donors with divergent philosophies are to be found implementing a patchwork of rural projects, which a bureaucracy lacking the necessary skills, manpower and finances is expected to coordinate. Although the aim is replication and, eventually, country-wide coverage, each project remains in splendid, costly isolation, and the poor recipient countries are increasingly unable to afford the follow-up costs.

l) Integrated planning and coordinated implementation. The closely linked causes of rural poverty and backwardness call for a comprehensive approach, but not necessarily an implementing structure that mirrors integrated planning. Rural development should be the subject of integrated, i.e. multi-sectoral, planning, but implementation should be sectorally coordinated by the relevant administrative bodies.

m) Open planning encourages development processes. While it is still undisputed that blueprint planning is very suitable for the planning of clearly defined technical projects, such as the improvement of infrastructure, rural development should be seen as an on-going interactive process planned openly (rolling planning).

Development policy in the past 40 years proves beyond any reasonable doubt that programmes for sustainable rural development and poverty alleviation require an enabling macroeconomic and policy environment. It tends to be a truism in this context to say that economic growth in low-income countries is essential if poverty programmes are to be launched with any chance of success. However, economic growth is not enough on its own.

What is needed is (a) qualitative growth geared to the needs and options of the poor target groups, (b) widely available social services, such as primary education (functional literacy) and preventive health care (including family planning services) and (c) decentralized economic and administrative infrastructure to create an effective framework for rural development programmes.

Growth policies aimed at specific target groups place the emphasis on agricultural, rural and small-scale industrial development, which is typically labour-intensive. The improvement of physical, social and administrative infrastructure should be geared primarily to the needs of the majority of the population.

A further important dimension for successful rural development is a political system that does not obstruct the self-organization of the poor and their representation at grass-roots level and develops the political will to see poverty as a problem for society as a whole.

# II Rural Development under the Conditions of the 1990s

## 6 Laying Suitable Foundations for Effective Poverty Orientation in Rural Development

Rural development can succeed if the economic and social environment is conducive to the comprehensive mobilization of potential that has hitherto lain dormant or been inadequately tapped. How can this be achieved? Where should development cooperation begin?

### Reform policies require managerial competence and administrative capacities

The illusion in the early years of the structural adjustment programmes that a significant adjustment of fundamental economic parameters and the restriction of government activities to core areas were enough to rehabilitate economies has given way to the realization that adjustment processes should be designed for the longer term and that they also require more profound sectoral structural adjustments; this is particularly true of agriculture, where the structural distortions need to be rectified.

It is now generally agreed that reforms of pricing and market policies are not enough on their own to solve the agricultural growth and social problems of structurally weak countries from the supply side. It is particularly clear from structural adjustment programmes (i.e. the prescribed liberalization of markets, the reduction of government subsidies and the raising of agricultural producer prices and of currency devaluations) in the poor countries of Africa that social destabilization may become more deeply rooted without sustained growth of production occurring. The accompanying political destabilization may be accepted in countries with a dictatorial bent, but a reform-induced deterioration of the situation of the poor majority of the population is unacceptable.

Fundamental structural constraints on the supply side usually need to be removed before prices can play their part as incentive and investment signals. The watchwords here include land reform, technology development, financial system, primary education and primary health care.

In these sectors complex reform programmes must be initiated and implemented, and for this the administrations of many developing countries have neither the competence nor the capacities. Even if government activities are successfully restricted to core tasks, as the structural adjustment programmes prescribe, government institutions are likely to be overextended unless they receive additional external support in the shape of manpower and funds.

In view of the limited prospects of success, however, attempts to overcome structural constraints with individual projects restricted to specific regions, a common practice in the past, should be made only where the political environment is favourable. Structural policy should rather be pursued with the instruments appropriate to it; individual projects can neither replace such instruments nor compel their use. However, this in no way signifies a radical rejection of individual projects organized at regional level, which may be a highly suitable means of implementing structural policy reform approaches. Technical cooperation thus faces new tasks, complementing conventional project aid with competent professional advice, combined with financial contributions towards the launching of (sectoral) reform programmes that encourage or even make possible the required structural change.

## Rural development needs effective local self-administration

In many developing countries distrust characterizes the relationship between centralistically structured public administration and the people. They expect little good of it, but depend on its often irregular, qualitatively questionable services, there being no private alternatives. In some cases services promised by law are obtainable only on payment of bribes and after several times of asking. Administrations are regarded as the self-service stores of the influential and powerful, who appear to have first call on government services.

Yet a high degree of frustration prevails among government officials too. This is due, for example, to low salaries, which are seldom paid on time, difficult living conditions at sometimes isolated locations, leading to the separation of families (to enable children to go to school), nepotism, paltry programme budgets and unsuitable technical equipment, if any. A very basic reason for the poor functioning of public administration is the limited decision-making authority delegated to local officials by central government; as a result lines of decision are long and ministries are completely overloaded.

The structural principle of centralist administration is monopoly of power and control rather than delegation of responsibility.

Examples show that rural communities and small rural towns too certainly have a potential for self-administration that is capable of development.[46] They are able to plan and implement development projects successfully; they are also able, if the law allows, to mobilize and organize local self-help in the form of contributions of money and labour.

Experience has also shown that the rural population is quite willing, within the limits of its financial possibilities, to pay for public services and the use of infrastructure, but expects socially just rules on the collection of taxes or levies and a transparent system of expenditure.

Structural adjustment programmes appear in this context to be a useful means of launching the necessary reforms of the administration and of achieving the necessary adjustment of the financial system.

### Rural development needs scope for self-help and self-organization

Willingness to take the initiative and the ability to organize self-help largely depend on society's room for manoeuvre, which in many cases is very limited. As the aim of self-help is to overcome the structural causes of poverty and to achieve a lasting improvement of physical and social living conditions, self-help movements easily clash with political forces that are afraid of losing power and influence. The task for development cooperation is to use political and financial instruments in such a way that the scope needed for self-help movements is created or safeguarded. Willingness to undertake reforms should be rewarded with development policy support of an appropriate quantity and quality and accompanied by a continuing policy dialogue, conducted not only with governments but with all relevant social groups.

Countries willing to undertake reforms should be offered fairly longterm assistance with the conceptualization, implementation and financing of sectoral policies to benefit poor target groups.

## 7 Target-group-oriented Sectoral Programmes as Elements of Comprehensive Rural Development

In addition to the creation of a beneficial political framework that provides the necessary scope for self-organization and self-help, there is a need for appropriate sectoral policies that help to eliminate the structural shortcomings which are the main cause of mass rural poverty and prevent it from being overcome. Areas of initial strategic importance, on which poverty-oriented development cooperation in favour of rural areas should focus, are:

- land and tenancy reforms,
- rural financial systems,
- basic social services and security systems.

## Land and tenancy reforms

As the unequal distribution of land and land use rights is a fundamental structural cause of rural poverty, comprehensive reforms of land ownership rights are clearly needed. Radical land reforms are, however, the exception and normally the outcome of revolutions or colonization rather than a consistent reform policy. As economic power and political power are usually all too closely linked, extensive land reforms regularly fail under the prevailing social conditions because of open opposition or delaying tactics on the part of landowners. Thus, exceptions and minor government redistribution measures aside, everything stays as it is.[47]

Land redistribution is not only desirable on social grounds but above all advantageous for macroeconomic development because of the stimulating effect it has on intensity of use, the development of productivity per unit of area and the scale of agricultural employment. However, advantages can be derived from small-scale farming only if the available technologies feature a high degree of divisibility of the factors of production used and if the latter are also traded in efficient factor markets. It should also be remembered that, where tenancy relationships are dissolved and redistribution occurs, all the production risks pass to the small farmers, who have no reserves with which to offset them. In view of the obvious investment and financing risks to which land reform policies expose the beneficiaries the effectiveness of the social safety net is likely to determine the pace of agricultural change.

In the light of market imperfections and the high risks inherent in production and marketing the present tenancy systems,[48] which are a typical consequence of the high concentration of land ownership, do not in any way appear to be obsolete social institutions; this may also explain their surprising stability. It is also true to say, however, that on both economic grounds (e.g. longer-term, land-improving innovations are impeded) and social grounds (e.g. the uncertainty of tenancies) tenancy systems, too, should eventually be ousted. Yet: "(Sharecropping) will continue to be prevalent as long as farms are small and farmers are poor, markets are underdeveloped, and infrastructure is weak in rural areas. It is poverty that leads to sharecropping, and not sharecropping that causes poverty."[49]

Tenancy reforms that seek to give tenants more security, to limit rents and to distribute costs more fairly between owners and tenants often prove to be politically and economically impossible to achieve. As long as the economically weak in the village are unable to express themselves freely because they are too vulnerable as individuals and their self-organization is unsuccessful, it will be very difficult to implement tenancy reform laws. An added factor is that - as striking examples in Africa and Asia show - technological change has a major influence on the pace and direction of changes in land tenure systems. Laws have difficulty resisting this pressure for change. It seems more important for tenants and landless farmers to be able to organize themselves in order the more effectively to safeguard their own rights and interests.

Besides legislation to protect tenants' fundamental rights and the promotion of organizations of farmers and the landless, poverty-oriented structural policy, which should be supported by development cooperation, requires above all else a set of promotional instruments which comply with market principles and encourage the voluntary sale of land ownership rights to those who actually work the land, i.e. small farmers. The provision by a land reform bank or an agricultural development bank of long-term, possibly even interest-free credits for tenants wanting to buy land might form the financial framework. To this end mortgage banks should be assisted with development cooperation funds. Landowners should be encouraged to sell with offers of additional, project-based investment funds and advice once sales have been completed. To make it easier for tenants to decide to buy land, special programmes of training and advice should be set up to prepare them for their new role as independent farmers.

The feudalistic land tenure systems of Asia and Latin America, where the division of inherited rights of land ownership and use can be achieved with legal and financial incentives, differ from the traditional communal or tribal agricultural systems of Africa, for example, in that the retention of land use entitlements usually requires some (minimal) farming of the land. The still significant subsistence mode of production and the prevalent division of farm labour between the sexes, in which women take primary responsibility for food production, may help to cause appreciable social and economic destabilization in African countries if the land tenure system is modernized to encourage individual property and ownership rights. Wherever the introduction of irrigation systems significantly increases the value of land, individualized land rights lead to major social distortions. For many African farmers of both sexes traditionally secured land use rights form part of a complex system of social security, the workings of which must be understood before any modernization is undertaken.

## Rural financial systems

Rural producers, i.e. small farmers, landless farmers, craftsmen, small traders, etc., need unbureaucratic and reliable access to financial resources to finance their inputs, to meet their subsistence requirements before the harvest, to make investments, to effect business transactions and to fund unusual requirements (weddings, illnesses, deaths).

Efforts under development policies to install rural financing systems long focused on the creation of specialized credit institutions, such as agricultural credit cooperatives and agricultural banks. They were intended to serve three purposes:

- to provide the credit needed by rural producers' for investment,
- to enable small farmers and producers to use modern, efficient technologies by keeping interest rates low and
- to counteract the informal private rural financial market (private money lenders), where interest rates were considered extortionate.

Hardly any of these ambitious goals was, however, achieved. It was found that many of the newly established agricultural and cooperative banks could not survive without continuous subsidies; the system of agricultural credit cooperatives began to ail and largely abandoned the rural credit business. As government imposed an upper limit on interest rates, the interest margin was nowhere wide enough for the financial institutions to perform the task assigned to them of lending to small and very small rural enterprises and cover their costs. In addition, the repayment rate, especially on agricultural production credits, was often less than 50 % of the funds lent.[50] This was due to deficient credit evaluation, poor credit surveillance, the inadequacy of advice given to inexperienced borrowers, the commercial failure of investments and poor repayment discipline among borrowers, who regarded credits provided by government as non-repayable transfers.

The majority of small and very small rural enterprises still have no access to institutional credit (estimates vary between 5 % in Africa and 15 % in Asia and Latin America).[51] Interest subsidies primarily benefited the economically strong, whereas the economically weak tended to be encouraged to make risky investments, over which they had too little control and on whose failure they were threatened with overindebtedness and social relegation.

In the majority of poor countries rural demand for financial services continues to be met by private money lenders, traders, landowners, pawnbrokers and neighbours and by such financial self-help organizations as credit and savings associations. As a rule interest rates in these usually unorganized rural financial markets are high;

empirically determined rates vary from 15 to 150 % p.a., with 30 to 40 % regarded as quite normal.[52] The high interest rates reflect, on the one hand, the price of capital in real terms and, on the other hand, the high credit risks where collateral is not provided, the cost of raising money and collecting information, the opportunity cost of capital and the financier's position in the local market. Furthermore, in the real world of small farmers, the landless and very small producers the average sums borrowed are minuscule and the credit period often very short; the economically logical consequence is that the negative scale effects are also reflected in the cost of credit.

Despite the high cost of borrowing, demand exceeds supply in some cases. This shows that it is not only the interest rate but also the availability of and continued access[53] to financial resources that determine the demand for credit among would-be borrowers without bank accounts. The small size of the typical rural credit transaction and the high frequency of turnover indicate the need for financial institutions which react flexibly, whose costs per transaction are low and which can be financed from the given interest margin. The typical bank in the formal sector can hardly meet these requirements: it works to fairly rigid rules on credit evaluation and lending, and its cost structure tends to favour the handling of comparatively large loans.[54] At present only decentralized financial institutions of the informal sector are likely to be able to satisfy the atomistically structured rural demand for credit. In many poor countries it will be some considerable time before the attainable turnover is sufficient to justify economically a network of village branches of banks.

Many income-generating initiatives taken by rural producers are thwarted by structural obstacles to borrowing. The relative importance of interest rates as a determinant of aggregate demand for credit cannot conceal the fact that they are the decisive criterion that leads to exclusion in certain cases. The absence of competition in village financial markets is the reason for many of the reports of excessively high interest rates. As the system of government-controlled credit institutions has been able to do little to remedy the situation, it is important for those affected to organize themselves.

For a long time the poor rural producer was seen as someone who, on the one hand, has a seemingly insatiable need for credit and, on the other hand, is neither able nor willing to save. The saving that was necessary for the economy took the form of compulsory saving through prices and taxes, while the few attempts to encourage voluntary saving proved relatively disappointing. Small farmers viewed formal banks and cooperatives with caution, if not suspicion. The autochthonous forms of financial self-help which had long since formed in all developing regions, such as the many traditional savings and credit associations (rotating savings and loan associations, tontines, savings clubs, etc.), were overlooked, or their social or economic relevance was underrated.[55] The dominant view now appears to be swinging to the other extreme, since it is considered to be almost beyond doubt that even the poorest population groups have savings, although they can be mobilized only if conditions are favourable.[56]

However, the lives of many people in the abjectly poor regions of numerous regions are probably so precarious that savings exist neither in kind nor as money and are hardly likely to be formed without encouragement from outside. Although here again it is true to say that saving may be an important form of self-help, the capacity for self-help which, say, the landless or virtually landless farmers potentially have must be strengthened before thoughts can turn to the formation of savings. Saving is then the second step and is taken to safeguard what has already been achieved.[57]

In ecologically fragile areas where harvests periodically fail all savings are used up during production and food crises, and in particularly bad cases massive disinvestment occurs, i.e. capital is consumed and destroyed. In such crises the traditional self-help mechanisms also fail, since the majority of the members of community-based safety nets are equally affected. Vertical and horizontal linkages between self-help groups and self-help networks may delay the failure of the support system, but the savings funds that can be mobilized impose fairly tight limits. Forging an institutional link between the informal savings and credit system and the formal banking system may lead to a lasting expansion of the available credit volume by giving access to refinancing funds, it will have the effect of reducing the interest rates charged by private money lenders by increasing competition, and it will make self-help mechanisms more crisis-proof. However, the risks inherent in easy access to financial resources must not be overlooked; what is needed is a set of transparent rules, linking credit creation to strict financial discipline and the successful mobilization of savings.

Wherever it has been possible to provide financial services that meet the needs of those who are traditionally ineligible for bank accounts, the consequent additional incomes and growth of self-confidence have led to a sustained improvement in the self-help capacity of the beneficiaries.

The successful cases[58] of self-help-oriented financing systems have a number of constituent and strategic factors in common, suggesting that these should be seen as a guide for future promotional programmes:

- The linking of saving and credit: continued access to credit should depend not only on the punctual and full repayment of initial loans but also on regular saving. Funds saved should be used as reserves and for securing future loans. The volume of new loans should depend on the amount saved.
- Group liability and social control: collateral for loans is replaced with the joint liability of credit groups; on the one hand, the groups exercise social control, on the other, they form the basis for joint productive and social efforts. Credit groups may be the nucleus of the more comprehensive self-organization of poor target groups[59] and may also act as intermediaries between the people and government in dealings with the administration and government service institutions. - The formation of small homogeneous, self-organized groups: a wide range of experience with credit groups shows that small (7-15 members) and socially homogeneous neighbourhood groups have the best chance of succeeding. It is important that the members' interests should be

the same in most respects, given their similar social and economic status, and that they should be able to learn from each other. Examples in Asia and Africa prove that groups of women have a better chance of succeeding than groups of men.

- Positive real interest rates: economically independent survival and the sustainability of financial self-help organizations and of formal financial institutions can be guaranteed only if the cost of providing credit (refinancing costs) and the cost of lending (transaction costs, credit risks and the opportunity cost of capital) are met, i.e. decapitalization is avoided.

- Meeting real transaction costs: both formal and informal financial institutions must have the legal means of recouping their transaction costs by pricing their products as they see fit; in the provision of very small credits for poor target groups banks will not as a rule be able to compete with self-administered financial self-help groups because their transaction costs are structurally higher.

- Unrestricted use of credit: to specify the only purposes for which credit may be used is to ignore the actual needs of small rural producers; they do not divide their lives into different spheres in accordance with the laws of business management; the lines separating consumption from production are fluid and indistinct. The example of the Grameen Bank also shows that bankers are seldom imaginative enough to prophesy all the investment opportunities recognized and seized by poor rural groups.

- Consideration of risks and efficiency: the success of poverty-oriented savings and credit programmes very much depends on a realistic assessment of the credit risks and of the economic efficiency of the target groups. Repayment methods and periods must take account of this, net savings possibly being a useful guide for assessing efficiency.[60]

- Combining financing, education and advice: while it is for the lending institutions to give financial advice, the social, technological and business advice that is needed should be arranged and provided by government service institutions outside the financing programmes. Credit groups facilitate the organizational task in this context, since delegated group members can act as multipliers.

- Confidence-building, community-oriented measures during the establishment phase: encouraging communal self-help can help to reduce the reluctance of marginal social groups to help themselves individually and communally, for which they are criticized everywhere. The provision of matching grants from development cooperation funds to supplement the community's efforts, both financial and in kind, to improve the social and productive infrastructure of villages might provide the initial spark needed to set a more sustained process of self-help in motion.

- Progressive graduation of groups: in many countries there is above all a need for the recognition and promotion of financial self-help groups that have formed without external support. In many ways such groups are already conducting themselves and abiding by rules that are essential if there is to be sustained financial self-help. After a relatively short period of assistance these groups can be formed into networks which are then gradually entrusted with the tasks of raising funds and forging links with the formal sector.[61] Where autochthonous financial self-help groups first have to be encouraged to form, it has been found advisable for groups to set up their own savings funds after an establishment phase - in much the same way as under the traditional ROSCA system - to promote learning processes and build confidence among the members, money then being lent within the group in accordance with rules it has itself laid down. Only when they have succeeded in mastering these procedures should financial self-help groups be linked to the formal system.[62]

Despite all the success undeniably achieved in promoting financial self-help initiatives, it would be irresponsible to overestimate the effectiveness of financial approaches as means of alleviating poverty through self-help: besides the extremely limited saving capacity of the members of the bottom 20 % of the rural income pyramid, social disadvantages - financial self-help groups too mirror the social stratification of the village - and individual disadvantages - poor education and training, few marketable skills - restrict the economic prospects of the absolutely poor and thus the impact of financial instruments.

## Basic social services and security systems

Structural reforms undertaken at macro and sectoral level with a view to strengthening the social dimension of regulatory and economic policy to the benefit of particularly underprivileged sections of the population contribute to the necessary creation of or increase in the scope for self-help and so present new options for self-determined economic development. Yet, given the dramatic dynamism of structural change in the developing countries, which is due to population growth and the increasing scarcity of resources and leads to social destabilization, these efforts to achieve stabilization by means that comply with market principles are unlikely to be enough in themselves for rural mass poverty to be overcome on any significant scale.[63] These reform programmes are more likely to lead to further and, to some extent, unavoidable social differentiation within poor population groups.[64]

One end of the social spectrum will again consist of groups which succeed in gaining or retaining control over sufficient means of production - whether they are farmers, craftsmen or small traders - to achieve a minimum standard of living. In the middle range, which is difficult to define accurately, a group tending to grow in size is likely to earn its living and so attain a certain degree of social security by offering its labour in the usually small and complex domestic labour markets or by seeking its fortune in temporary emigration. The other end of the spectrum will consist of those who cannot earn a living because of sickness, age or mental or physical disability or depend on the provision of private and/or public social (welfare) services at times of general crisis and personal emergency.

Even the fortunate owners of means of production that yield enough for them to make a modest living in normal circumstances may suffer a temporary or permanent loss of productive capacity due to failed harvests, losses of revenue, sickness, invalidity, etc., which threatens their standard of living. Where a loss of productive capacity is no more than temporary, however, they may succeed in safeguarding their standard of living for a time by saving or borrowing.

For wage-earners without productive resources of their own, especially those employed in agriculture and the informal sector, the danger of losing jobs and thus their income is incomparably greater. At the low wages they receive, they have very little prospect of saving enough to live on in an emergency; nor, as a rule, are they in a position to borrow. They depend on the social safety net to survive periods of unemployment. Those who are incapable of self-help, particularly widows, orphans, the elderly, invalids and the disabled, depend on this safety net for their survival from the outset.

In a farming society the social safety net, i.e. the social security system, to which the rural population can usually resort to avert emergency situations is largely based on family relationships and reciprocal exchanges at village level. These traditional social security systems are, however, showing signs of disintegrating because of social change and a deterioration of coping capacity due to persistent economic crises. Although family ties remain the pivot on which traditional social security turns, processes of impoverishment are weakening the economic capacity of many families and helping to make them more dependent on outside help.

The recurrent decline in production in Africa's Sahel region is a frightening example of the increasingly frequent failure of family and communal self-help and survival strategies due to the fact that productive capacities have dwindled to the level where it has no longer been possible to accumulate sufficient surpluses for the next crisis.[65] Many families are today simply unable to give ailing relatives the assistance they need. How far poverty-induced migration weakens or strengthens the self-help capacity of families in emergency situations varies from case to case.

The forms of traditional solidarity commonly found in farming societies offer only a very limited defence against frequently recurring emergencies, since the organizational and financial capacity of the communities concerned is weak and their operational rules are increasingly geared to the principle of equivalence rather than the idea of acting in solidarity in emergency situations.[66] This assessment, reported from West Africa, concerns both traditional mutual assistance associations and tontines (savings and credit associations for the financing of ceremonies, especially funerals). Governments and donors thus face enormous social tasks, extending well beyond the mitigation of any social hardship caused by reform processes; there is a need both for widespread basic social services that strengthen the capacity for self-help and for an appropriate system of social security that so combines informal and formal elements that an effective and also affordable social safety net emerges. In this process the further destabilization and eventual destruction of traditional security strategies and safety nets must be avoided.

## Basic social services

The outstanding role that the provision of effective basic social services, i.e. primary health care and primary education, can play in poverty alleviation and the achievement of a higher level of social security is evident from Sri Lanka and the Indian state of Kerala. Both have been extremely successful in the past in creating more social security and a better quality of life for the majority of the population.[67] Many African examples, on the other hand, show that the absence of effective basic social services limits and eventually erodes the capacity for self-help: poor health, little knowledge of the links between the environment and health, the absence of marketable skills, seriously impeded access to modern information, etc. reduce the coping capacity.

The development and provision of basic social services have always been important components of rural development projects; as a rule they complement and parallel the few, poor-quality government services. The resulting continuing need for grants to enable schools, health posts, etc. to remain in operation easily exceeds the funds available to the sponsors in most cases. They then react by failing to undertake necessary repairs and maintenance work and/or by reducing the services provided. This is essentially the outcome of two mistakes that need to be corrected:

- For construction and technical equipment donors and recipients, in unholy alliance, have often chosen unrealistic and alien standards, entailing higher capital and follow-up costs than would have been incurred if local standards had been more successfully adapted. Furthermore, yearning for technical modernity encourages relatively pronounced centrality, making self-administration difficult and putting the rural poor at a serious disadvantage.
- Welfare-statist thinking among the donors has also encouraged approaches that have sought to ensure free access to social services for everyone, but have in effect primarily subsidized those who have always been able to meet the opportunity cost of non-practical general education and the high cost of access to health posts and clinics.

Experience has shown that a high geographical density of social services must be provided at a decentralized level if the social and economic barriers to access for the poor are to be lowered. Such basic infrastructure as buildings can be constructed by local self-help groups with their own labour. While the principle of free access to health care and education should be upheld in the case of basic local services, realistic charges, i.e. ones that cover recurrent costs, should be made for services at a higher level of centrality, with provision made for social discounts, despite the difficulty of proving the need for reductions. The necessary country-wide provision of basic social services cannot be achieved with isolated projects that are run by different donor organizations and are difficult to coordinate. In this case project aid should give way to sectorally oriented programme aid.

## Social security systems

Serious risks against which people cannot adequately protect themselves threaten the normal pattern of life. Besides the general hazards of life, it is above all production risks that may give rise to emergency situations. Although individual provision for risks is very important, the greater the uncertainty about the occurrence of losses, their scale and duration, the more immediate the limits to which such provision is subject.[68] It is here that systems of social security that make provision for risks at the level of a community or society step in. Social security means both prevention, i.e. avoiding a risk, and alleviating the consequences of an emergency situation caused by a risk that could not be averted. As a task for social security, precluding risks still seems very important in many rural societies; functional basic social services also help by providing preventive social security.

As farming societies face high and often unpredictable risks, however, it is primarily the effective alleviation of the consequences of losses, i.e. the general prevention of an emergency situation or a lasting deterioration of living conditions, that forms the core of an appropriate social security system. In principle this can be achieved in three ways: intertemporal, interregional and interpersonal exchanges and redis-tribution. While, for example, the immediate consequences of a partial loss of harvest can be mitigated where individuals or communities lay in (emergency) stocks, a continuing drought soon leads to the depletion of local stocks in poor farming societies and so exhausts a local system of food security in a relatively short time. Interregional redistribution may help in such cases, but effective communication and transport is then extremely important.

The most important form of social security is based on interpersonal redistribution in communities that share risks and act in solidarity. The potential for social security within groups depends on a workable relationship between producers and non-producers. Whether the required spreading of risks succeeds depends on their nature, on the size, composition and sustainability of the group and on the redistribution principle chosen.[69] The principle of equivalence (insurance) and the principle of solidarity are the two poles between which redistribution actually occurs.

The sustained strengthening of existing family and communal security systems is likely to be one of the major tasks for development cooperation that has wide-ranging alleviation of poverty as its overriding objective. As the traditional security systems are increasingly unable to ensure that risks are comprehensively spread because their resources are limited and social destabilization is growing, ways must be sought to create a robust and flexible social security network by linking (informal) self-help efforts on the one hand and combining them with (formal) cooperative, occupational and government social security provisions on the other. But how to take advantage of the benefits of local, self-administered support structures that consist in the rapid acquisition of relatively complete information, a transparent and needs-oriented arrangement and easy monitoring to prevent abuses, while avoiding the disadvantages of small organizations that tend to be financially weak? There is much to be said for a multi-tier system in which advantage can be taken of the opportunities at the higher levels for raising funds and obtaining grants, but which adheres to the principle of local self-administration. A multi-tier system of social security must include not only elements of social insurance (health, accident and invalidity insurance) but also social assistance, so that those who are permanently or temporarily unable to fend for themselves may be sure of a minimum standard of living.

Besides effective risk-spreading to ward off emergency situations due to the general hazards of life, a sustainable increase in the capacity for self-help is vital if there is to be an adequate degree of social security. More employment and fewer production risks are important action parameters in this context. More employment in the form of cash for work under local and regional development programmes will provide the incomes that may help to restabilize local security systems; equally, protection against unpredictable production risks, and especially the securing of production and general investment credits by guarantee funds, may make a significant contribution to increased social security.

The major advantages of local or regional employment programmes on a cash for work basis are that, with the aid of calculated decisions on wage rates and the methods and management principles to be applied, those employed can be chosen to suit the required target group (self-selecting social security scheme). The advantages to the public client are also obvious: not only does he avoid transfers to poor groups, but the actual capital cost of self-administered infrastructure projects is considerably lower than that of comparable projects put out to commercial tender.

The sustained linking of institutional credit and collateral should succeed where the provision of group credit is accompanied by effective social control and the guarantee fund, initially fed from outside, gradually gives way to saving within the group. The collateral insurance of harvests and livestock, on the other hand, is hardly likely to come into general use as a means of increasing social security because of insurance and institutional problems and the difficulty of preventing abuse.

## 8 Necessary Modification of the Project Approaches

The effectiveness of individual projects that seek to bring about sustained and wide-ranging social, institutional and technological change increasingly depends on coherent sectoral and macro policies to create a conducive basis. If these policies are not coherent, projects operate with high frictional losses, tie up many scarce resources unnecessarily and all too often fail to set dynamic processes in motion, leading at best to reasonably effective physical structures and isolated increases in productivity. The great disillusionment when success does not materialize also depresses the helpers.

Many of the development experts assigned, who appear with the self-confident claim to be the bearers of knowledge relevant to development and, as catalytic change agents, increasingly seek to enable their counterparts to do what is right for the target groups at the right time, fail to make any progress, disillusioned by permanent frustration. They often take a cynical view of development, which allows them to maximize their private welfare without too guilty a conscience.[70]

On the other hand, there are the change agents who identify niches and gaps while working with grass-roots self-help organizations and help to disseminate exemplary institutional and technical innovations. They are usually closer in origin, thinking and acting to the communities they serve than the professional helpers assigned under international and multilateral development cooperation agreements.

After almost 40 years of manpower cooperation with the poor countries there is much to be said for a shift of emphasis from project-centred rural development to programme-centred political cooperation for rural areas. Changing from projects to programmes means above all abandoning pilot projects with a limited impact, their replication being largely unsuccessful because of structural factors. In contrast, programmed cooperation should mainly contribute to the structural transformation of rural society within the framework of general policy and with the support of society as a whole.

This reorientation of cooperation towards a comprehensive form of rural development will, on the one hand, result in the level of action being raised and, on the other hand, qualify the role played by individual rural projects, including RRD projects, which are restricted to a given region and subject to time limits.

Adjustments to project concepts, meaning their continuation in a more refined form, are unlikely to be enough, since raising the level of action also presupposes effective donor coordination on a far larger scale than in the past. This will be difficult to achieve if each bilateral donor applies relatively inflexible project concepts. The mixture of rural development approaches that has emerged in many places initially exceeded the coordination capacity of national administrations and eventually created a patchwork of approaches to promotion which were difficult to integrate and whose follow-up costs to the countries concerned are extremely high.

In the light of this experience the main tasks for rural development projects should now be:

- support for government administrations in the development of services geared to demand;
- concerted support for emerging social organizations and self-help movements;
- the strengthening of local self-administration by means of participatory, action-oriented planning;
- the promotion of rural-urban linkages;
- systematic mobilization and utilization of local know-how.

The main aim should be to create institutional, administrative, technological and manpower capacities (capacity-building) that enable existing human and natural resources to be used more effectively.

While most rural development projects have hitherto been implementing or advisory entities,[71] it is now felt that projects should be seen as intermediary structures, building bridges between government and people; this linking task requires the project not only to support the self-organization of those concerned but also the establishment or promotion of networked (communication) structures and willingness to play the role of the people's advocate (wide-ranging involvement of those concerned, transparency of decisions taken by the administration, the rule of law).

The vision of a new type of rural development project is incompatible with projects which are converted into auxiliary organs of a patrimonial state, which seek to satisfy their own clientele with the help of external funds and foreign personnel and in which striking a political and regional balance is usually more important in the decision to implement a project than criteria relating to potential and needs. Rural projects of the new type should be vehicles for the participatory, i.e. democratic, formulation and implementation of sectoral policies, i.e. the project and sectoral policies should be complementary and should together constitute the rural development programme.

At present development cooperation is not prepared for the new requirements either mentally or organizationally. Thinking in terms of individual projects easily leads to linkages being overlooked, but complies with a bureaucratic logic: administrative and fiscal segmentation into independent projects facilitates decision-making and spreads the risks wider. (Large-scale) technical projects with clearly defined physical outputs usually meet with less resistance than social projects that have institutional or organizational processes as their goals. They are more easily accepted, since it is assumed that they will lead to the desired economies of scale; but they are also preferred because their planners claim there is a clear and comprehensible link between inputs financed with development cooperation funds and the outputs that are sought (increase in productivity).[72] The bureaucratic logic of the development administration corresponds to the turnover logic of the implementing organizations.

If rural development is to be set in motion and to have a wide impact under the conditions of the 1990s, the supply structure of development cooperation should be so adjusted that policy advice at the meso level, technical and financial cooperation and interactive rural projects can be combined to form a programme package. This will mean abandoning the pastiche of unlinked agricultural and rural projects under the auspices of benevolent donors. The many individual projects of the donor community all too often add up to a whole that turns the statement "the whole is greater than the sum of its parts" upside down. The burdensome task of coordination is assigned to a notoriously inefficient administration in the recipient countries, of which it is asking too much to make a proper review of the many individual and usually short-term projects for their development policy and fiscal implications.

How to overcome this dilemma? Experience in some countries shows that a strategy of promoting rural areas that favours the emerging third sector (professional organizations, interest groups, NGOs, savings and credit associations, etc.) helps to relieve government of non-sovereign functions, which it is otherwise able to perform only at considerable expense and with limited economic and social effectiveness; at the same time viable self-help organizations and structures eventually emerge to form a new non-governmental partnership structure for development cooperation. Paraproject approaches based on decentralized or local initiatives and assisted directly with local grants or financial contributions (direct financing) are commensurate with this promotional strategy; financial assistance may be supplemented by offers of advice, which should be mobilized at local level as far as possible.[73] However, effective and needs-oriented cooperation with a country that is opening up, becoming democratic and developing decentralized decision-making structures also requires of the donors extensive local professional and administrative representation, which many of them do not have at present. A further thorny problem on the way to a new aid policy for rural areas is coordination among the various bilateral and multilateral donors. Having been desirable in the past, effective donor cooperation is essential in the age of structural adjustment programmes. It may be possible to organize a sectoral division of labour among donors in small countries, but coordinated participation by several donors is undoubtedly preferable on political grounds. In larger countries a sectoral division of labour is usually impossible if only because of the scale of the programme; donor consortia are needed in such cases. However, efficient coordination mechanisms are then required. A conceivable model would be one in which a lead agency headed the consortium for certain sectoral programmes.

At the end of a fairly long process the concerted efforts should then bear fruit, and there should be efficient implementing organizations and sufficient local capacities for rural development programmes to be planned, implemented and evaluated by countries on their own responsibility, with a minimum of external intervention.

# Notes

1 **U. Lele,** *The Design of Rural Development: Lessons from Africa*, Baltimore and London 1975, p. 123.

2 See, for example, **R. Chambers,** "Project Selection for Poverty - Focused Rural Development: Simple is Optimal", *World Development*, Vol. 6 (1978), No. 2, and **H. Ruthenberg,** "Is Integrated Rural Development a Suitable Approach to Attack Rural Poverty?", *Quarterly Journal of International Agriculture*, Vol. 20 (1981), No. 1.

3 **Federal Ministry for Economic Cooperation,** "Querschnittsanalyse von Projekten der 'ländlichen Regionalentwicklung'", Ref. 201, Bonn, June 1990.

4 **D.C. Korten,** "Organizing for Rural Development: A Learning Process", *Development Digest*, Vol. XX (1982), No. 2, p. 5.

5 **P.H. Coombs / M. Ahmed,** *Attacking Rural Poverty*, The Johns Hopkins University Press, Baltimore and London 1974, pp. 70 f., quoted in **J. Kumar,** *Integrated Rural Development. Perspectives and Prospects (1952 - 1982)*, Delhi 1987, p. 59.

6 **D.W. Jorgensen,** "The Development of a Dual Economy", *Economic Journal*, Vol. 71 (1961), pp. 309-334; **G. Ranis / J.C.H. Fei,** "A Theory of Economic Development", *American Economic Review*, Vol. 51 (1961), pp. 252-273; **B. Johnston / J.W. Mellor,** "The Role of Agriculture in Economic Development", *American Economic Review*, Vol. 51 (1961), pp. 566-593.

7 **T.W. Schultz,** *Transforming Traditional Agriculture*, New Haven 1964.

8 The new status enjoyed by agricultural research at international level is also underlined by the establishment in 1971 of the Consultative Group on International Agricultural Research (CGIAR) under the joint auspices of the World Bank, UNDP and FAO as a financially and programmatically powerful international agricultural research organization.

9 For the wide range of social effects of the Green Revolution see **I. Singh,** *The Great Ascent. The Rural Poor in South Asia*, Baltimore and London 1990.

10 See **J. Kumar,** *Integrated Rural* ..., op. cit., pp. 68 ff.

11 **World Bank,** "Rural Development", Sector Policy Paper, Washington D.C., Feb. 1975.

12 **World Bank,** "Rural Development. The World Bank Experience, 1965 - 1986", Washington D.C. 1988, p. xiv.

13 **Ibid.,** pp. xiv f.

14 **M. Yudelman,** "Agricultural Development in the Third World: The World Bank Exerperience, Social Strategies," Forschungsberichte, Vol. 1 (1986), No. 3, Soziologisches Seminar der Universität Basel, p. 30.

15 See **Y. Hayami / V. Ruttan,** *Agricultural Development*, Baltimore 1971; **B.F. Johnston / P. Kilby,** *Agriculture and Structural Transformation. Economic Strategies in Late-Developing Countries*, New York, London and Toronto 1975; **J.W. Mellor,** *The New Economies of Growth*, Ithaca 1976.

16 **H.B. Chenery et al.,** *Redistribution with Growth*, London 1974.

17 **International Labour Office,** *Employment, Growth, and Basic Needs. A One World Problem*, Geneva 1976.

18 See **German Development Institute,** *Grundbedürfnisorientierte ländliche Entwicklung*, Berlin 1980, therein **P.P. Waller,** "Das Grundbedürfniskonzept und seine Umsetzung in der entwicklungspolitischen Praxis", pp. 1 ff.

19 An "Integrated Rural Development Programme" was launched in East Pakistan in 1970 for, among other things, the institutional implementation of the Comilla approach. An international symposium on "Agricultural Institutions for Integrated Rural Development" was held by the FAO and Sweden's SIDA as early as 1971.

20 See, for example, **H. Dequin,** "Neue Genossenschaften im Programm der integrierten Entwicklung in Bangladesh", *Zeitschrift für Ausländische Landwirtschaft*, Vol. 14 (1975), pp. 159-163.

21 **L.E. Birgegard,** "A Review of Experiences with Integrated Rural Development", *Manchester Papers on Development*, Vol. IV, No. 1 (1988), pp. 4-27.

22 See **V. Steigerwald,** "Systemorientierung in der integrierten ländlichen Entwicklung", *Studien zur Integrierten Ländlichen Entwicklung*, Vol. 31, Hamburg 1989, pp. 9-38.

23 **M. Yudelman,** "Die Rolle der Landwirtschaft bei Projekten der integrierten ländlichen Entwicklung. Die Erfahrungen der Weltbank", in: T. Dams (ed.), *Integrierte ländliche Entwicklung*, Munich, Mainz 1980, pp. 120-138.

24 See, for example, **H. Ruthenberg,** "Rural Development - eine Sackgasse?", *Zeitschrift für Ausländische Landwirtschaft*, Vol. 14 (1975), pp. 3 f.

25 **K.M. Fischer et al.,** *Ländliche Entwicklung. Ein Leitfaden zur Konzeption, Planung und Durchführung armutsorientierter Entwicklungsprojekte*, research commissioned by the Federal Ministry for Economic Cooperation, Hamburg 1978, pp. 1 f.

26 While the term "integrated rural development" (IRD) has become internationally accepted, "ländliche Entwicklung" (rural development) or "ländliche Regionalentwicklung" (LRE) (regional rural development) is the term officially used in the Federal Republic of Germany, LRE standing for a

specific type of German technical cooperation project and "ländliche Entwicklung" for an multisectoral development cooperation concept.

27 **Federal Ministry for Economic Cooperation (BMZ),** *Konzept zur Förderung der Ländlichen Entwicklung. Grundsätze für Planung und Durchführung von Vorhaben zur Entwicklung des Ländlichen Raumes*, Bonn, July 1988.

28 **Ibid.,** pp. 5 f.

29 For example, **D. Senghaas,** "Integrierte ländliche Entwicklung", *epd - Entwicklungspolitik*, No. 1 (1981), pp. 23-25.

30 **H. Ruthenberg,** "Is Integrated Rural Development a Suitable Approach to Attack Rural Poverty?", *Quarterly Journal of International Agriculture*, Vol. 20, No. 1 (1981), p. 13.

31 **T. Dams et al.,** "Integrierte ländliche Entwicklung - Theorie, Konzepte, Erfahrungen, Programme", *Studien zur Ländlichen Entwicklung*, Vol. 16, Hamburg 1985, especially pp. 77 ff.; **V. Steigerwald,** op. cit.

32 **W. Manig,** *Integrierte rurale Entwicklung*, Kiel 1985.

33 **Ibid.,** p. 169; as a theoretical basis Manig proposes, on the one hand, Nurske's "balanced growth" strategy as the sectoral and time frame for decision-making and, on the other hand, Christaller's theory of central places as the spatial decision-making concept.

34 **World Bank,** "Rural Development", Sector Policy Paper, Washington D.C. 1975, pp. 60 f.

35 **World Bank,** *Rural Development. The World Bank Experience, 1965 - 1986*, Washington D.C. 1988, Annex 5, Table 1.

36 **K. Kumar,** "A.I.D.'s Experience with Integrated Rural Development Projects", A.I.D. Program Evaluation Report No. 19, Washington, July 1987.

37 **Bundesstelle für Entwicklungshilfe (BfE),** FlDeutsche Agrarhilfe - was? wo? wie?, Frankfurt 1972, p. 2.

38 **Ibid.,** p. 164.

39 **Ibid.,** p. 128.

40 **P.P. Waller,** *Grundbedürfnisorientierte Regionalentwicklung, GDI*, Berlin 1984.

41 **G. Elwert,** in a paper delivered at a DSE conference held in Tschortau in April 1991 on the subject of "Participatory Approaches in the Promotion of Self-Help Organizations".

42 **T. Rauch,** *LRE-aktuell*, "Strategieelemente für eine Umsetzung des LRE-Konzeptes unter veränderten Rahmenbedingungen", study commissioned by GTZ, April 1991.

43 See **GTZ (ed.),** *Where there is no participation*, Eschborn 1991, and **DSE,** "Participatory Approaches ...", op. cit., Workshop Report, April 1991.

44 **L.E. Birgegard,** *A Review of Experiences with Integrated Rural Development (IRD)*, Swedish University of Agricultural Sciences, International Rural Development Centre, RD Analysis Section, Uppsala, March 1987.

45 **H. Brandt / H. Lembke,** *Entwicklungshilfe als Dauerzuwendung am Beispiel der Sahelländer*, GDI, Berlin 1988.

46 **U. Otzen et al.,** *Development Management from Below*, GDI, Berlin 1988.

47 See **FAO**, *The Impact of Development Strategies on the Rural Poor*, WCARRD - Ten Years of Follow-Up, Rome 1988, pp. 33-47; **H. Gsänger,** *Perspektiven der Agrarreformpolitik Simbabwes im Lichte äthiopischer and kenianischer Erfahrungen*, GDI, Berlin 1985.

48 This is also true of share tenancy systems, which, unlike cash tenancy systems, where output is irrelevant, share the production and marketing risks proportionally between the owner and the tenant.

49 **I. Singh,** *The Great Ascent. The Rural Poor in South Asia*, Baltimore and London 1990, p. 287.

50 **H.J. Mittendorf / E. Kropp,** "Rural Financial Markets in Africa: The Challenge for Reform", *Entwicklung und ländlicher Raum*, No. 6 (1990), offprint by GTZ.

51 **J. von Stockhausen,** "Strukturwirksame Förderung der Agrarfinanzierung in Entwicklungsländern", *Agrarwirtschaft*, Vol. 40 (1991), No. 4, pp. 105-111.

52 **A.G. Chandavarkar,** "The Informal Financial Sector in Developing Countries: Analysis, Evidence, and Policy Implications", revised version of a paper prepared for the SEACAN Seminar on Unorganized Money Markets, Yogyakarta, Indonesia, 20 - 22 Nov. 1985, International Monetary Fund, Feb. 1986.

53 **T. Biggs et al.,** "On Minimalist Credit Programs", *Savings and Development*, Vol. 15, No. 1, pp. 39-52.

54 **F.J.A. Bouman,** "Informal Rural Finance. An Aladdin's Lamp of Information", *Sociologia Ruralis*, Vol. 30, No. 2 (1990), pp. 155-173.

55 Aspects of the propensity to save, how it can be influenced and the role that ROSCAs may play in this context were considered in: **C. Geertz,** "The Rotating Credit Association: A 'Middle Rung' in Development", *Economic Development and Cultural Change*, Vol. 11 (1962), pp. 241-163.

56 Gemeinsame Arbeitsgruppe "Armutsbekämpfung durch Selbsthilfe", Ergebnisbericht aus dem Arbeitsschwerpunkt "Sparen und Kredit" (AS 1/2), Bonn, April 1989.

57 A successful example of the promotion of financial self-help by using revolving credit funds entailing an obligation to save and administered by the beneficiaries themselves is the "Self-reliant Development of the Poor by the Poor" project now being implemented in four rural districts of Nepal by a Nepalese NGO, Helvetas and GTZ.

58 See, for example, **P. Egger,** "Banking for the Rural Poor: Lessons from Some Innovative Savings and Credit Schemes", *International Labour Review*, Vol. 125, No. 4 (1986), pp. 447-462; **D. Hulme,** "Can the Grameen Bank be Replicated? Recent experiments in Malaysia, Malawi and Sri Lanka", *Development Policy Review*, Vol. 8 (1990), pp. 287-300; **T. Biggs et al.,** "On Minimalist Credit Programs, Savings and Development", Vol. 15, No. 1 (1991), pp. 39-52.

59 Starting with credit groups, the "Dhading model" that forms part of the Small Farmer Development Project of Nepal's Agricultural Development Bank seeks to amalgamate basic groups into intermediary SHOs ("Intergroups"), which then form a small farmers' organization. Success so far (since 1989) has been promising; see **Human Resources Development Center,** SFDP/ Dhading, Briefing Book, Katmandu, May 1990 (unpublished).

60 **F. von Thun,** "Ansätze zur Armutsbekämpfung durch Selbsthilfe." Bericht über die internationale Tagung von BMZ und DSE im Januar 1985 in Feldafing, Feldafing 1985, p. 13.

61 For various approaches to the linking of informal and formal financial institutions see, for example, **GTZ/APRACA (eds),** *Linking Self-Help Groups and Banks in Developing Countries,* Eschborn 1989, and **H.D. Seidel,** "Microfinance for Microenterprises: Some Practical Experiences of Linkages between Formal and Informal Financial Institutions in Indonesia", paper for a symposium of the Royal Tropical Institute, Amsterdam, on "Sharing Poverty or Creating Wealth? Access to Credit for Women's Enterprises", Jan. 1991.

62 A good example of this strategy is GTZ's PAK-German Self-Help Project, Baluchistan/Pakistan.

63 With qualifications, this is also true of such sectoral projects that have been discussed as "promotion of the transfer of land ownership to the farmers" and financing systems geared to specific target groups.

64 This is confirmed by experience of the social effects of the structural adjustment programmes, especially in Africa; see **G. Lachenmann,** "Soziale Bewegungen als gesellschaftliche Kraft im Demokratisierungsprozess in Afrika?", *Afrika Spektrum*, Vol. 26, No. 1 (1991), pp. 73-97.

65 See **E. Ahmad,** "Social Security and the Poor. Choices for Developing Countries", *The World Bank Observer*, Vol. 6 (1991), No. 1, pp. 105-127.

66 See **R. Frey Nakonz,** "Solidarität und soziale Sicherung bei den Aizo (Südbenin)", unpublished manuscript, December 1990.

67 **S.R. Osmani,** "Social Security in South Asia", in: E. Ahmad / J. Dreze / J. Hills / A. Sen (eds), *Social Security in Developing Countries*, Oxford 1991, pp. 305-355.

68 See **M. Partsch,** *Prinzipien und Formen sozialer Sicherung in nicht-industriellen Gesellschaften*, Berlin 1983, and **H. Lampert,** *Lehrbuch der Sozialpolitik*, Berlin, Heidelberg, New York, Tokyo 1985.

69 See **M. Partsch,** *Prinzipien* ..., op. cit., p. 65.

70 A somewhat simplistic description of the world of the development experts, but one that accurately reflects the facts, can be found in: **G. Hancock,** *Händler der Armut*, Munich 1989.

71 See **T. Rauch,** LRE-aktuell. "Strategieelemente für eine Umsetzung des LRE-Konzeptes unter veränderten Rahmenbedingungen", study commissioned by GTZ, April 1991.

72 In the German ZOPP planning procedure this is clearly reflected in the vertical logic: if A, then B; if B, then at least a contribution to C. Aspects of the politically, fiscally and socially appropriate implementation strategy are discussed during the workshop breaks. The socio-political dicsussion on develoment takes place in the "assumption column of the PPM".

73 See **N. Uphoff,** "Paraprojects as New Modes of International Development Assistance", *World Development*, Vol. 18, No. 10 (1990), pp. 1401-1411.

# Bibliography

**Adams, D.W.**, "Rotating Savings and Credit Associations in Bolivia", *Savings and Development*, Vol. 13, 1989, No. 3, pp. 219-235

**Adams, D.W. / R.C. Vogel**, "Rural Financial Markets in Low-Income Countries. Recent Controversies and Lessons", *World Development*, Vol. 14, 1986, No. 4, pp. 477-487

**Adera, A.**, "Agricultural Credit and the Mobilization of Resources in Africa", *Savings and Development*, Vol. 11, 1987, No. 1, pp. 29-73

**Ahmed, Z.U.**, "Effective Costs of Rural Loans in Bangladesh", *World Development*, Vol. 17, 1989, No. 3, pp. 357-363

**Alibert, J.**, "Le cas original de tontines camerounaises, phénomène de société", *Marchés Tropicaux*, 17-24 August 1990, pp. 2375-2378

**Baldus, R.D. / C. Kohlbach / G. Ulrich**, Fachseminar "Selbsthilfe in der ländlichen Entwicklung", GTZ/DSE, Feldafing 1983

**Bari, F.**, *Small Efforts by Small Farmers. Attempt Towards Participatory Growth of Grass-Root Organisations*, Bangladesh Academy for Rural Development, Comilla 1987

**Bauer, G.**, "Hemmnisse bei der Umsetzung einer integrierten ländlichen Entwicklungspolitik in Tanzania", *Entwicklung und ländlicher Raum*, No. 1, 1983, pp. 22-24

**Bedard, G. (ed.)**, "Fighting Poverty through Self-Help", Report on the IIIrd Conference organised by the DSE, Feldafing 1989

-, "Saving and Credit as Instruments of Self-Reliant Development of the Poor", International Workshop organised by the DSE, Feldafing 1988

**Binnendijk, A.**, "A.I.D.'s Experience with Rural Development. Project-Specific Factors Affecting Performance", paper prepared for presentation at the Rural Development Seminar, World Bank, Paris Office, February 1988

**Birgegard, L.E.**, *A Review of Experiences with Integrated Rural Development (IRD)*, Issue Paper No. 3, International Rural Development Centre, Swedish University of Agricultural Sciences, Uppsala 1987

-, *A Modified Approach to Multi-Sectoral Area Development*, International Rural Development Centre, Swedish University of Agricultural Sciences, Uppsala 1987

-, "A Review of Experiences with Integrated Rural Development", Manchester Papers on Development, Vol. 4, 1988, No. 1, pp. 4-27

**Bouman, F.J.A.**, "Informal Rural Finance. An Aladdin's Lamp of Information", *Sociologia Ruralis*, Vol. 30, 1990, No. 2, pp. 155-173

**Brandt, H. / H.H. Lembke,** ***Entwicklungshilfe als Dauerzuwendung am Beispiel der Sahelländer***, GDI, Berlin 1988

**Brandt, V.S.R. / J.W. Cheong,** "Top-Down and Bottom-Up Rural Planning in South Korea", *Development Digest*, Vol. 20, 1982, No. 2, pp. 38-56

**Braunmühl, C. von (ed.),** *Fighting Poverty through Self-Help*, Report on the IInd Conference organised by the DSE, Feldafing 1987

**Braverman, A. / J.J. Guasch,** "Rural Credit Markets and Institutions in Developing Countries. Lessons for Policy Analysis from Practice and Modern Theory", *World Development*, Vol. 14, 1986, No. 10/11, pp. 1253-1267

**Bundesstelle für Entwicklungshilfe (BfE),** *Deutsche Agrarhilfe - was? wo? wie?*, Frankfurt 1972

**Burkett, P.,** "Group Lending Programs and Rural Finance in Developing Countries", *Savings and Development*, Vol. 13, 1989, No. 4, pp. 401-418

**Carroll, T.F.,** "Group Credit for Small Farmers", *Development Digest*, Vol. 12, 1974, No. 1, pp. 3-14

**Chambers, R.,** "Guiding Research Toward Technologies to Meet Regional Rural Needs", *Development Digest*, Vol. 20, 1982, No. 2, pp. 31-37

-, *Rural Development. Putting the Last First*, London, Lagos, New York 1982

-, "Project Selection for Poverty - Focused Rural Development: Simple is Optimal", *World Development*, Vol. 6, 1978, No. 2

**Chandavarkar, A.G.,** "The Non-Institutional Financial Sector in Developing Countries. Macroeconomic Implications for Savings Policies", *Savings and Development*, Vol. 9, 1985, No. 2, pp. 129-140

**Chao-Beroff, R. / C. Delhaye,** "Les Caisses Villageoises d'Epargne et de Crédit Autogérées", Compte-rendu de l'atelier international à Mopti, Mali 1989

**Chenery, H.B. et al.,** *Redistribution with Growth*, London 1974

**Conyers, D.,** "Future Directions in Development Studies: The Case of Decentralization", *World Development*, Vol. 14, 1986, No. 5, pp. 593-603

**Coombs, P.H. / M. Ahmed,** *Attacking Rural Poverty*, The Johns Hopkins University Press, Baltimore and London 1974

**Dams, T. / H. de Haen / H. Kötter / H.U. Timm / E. Zurek,** "Integrierte ländliche Entwicklung - Theorie, Konzepte, Erfahrungen, Programme", Studien zur Integrierten Ländlichen Entwicklung, Vol. 16, Hamburg 1985

**Dequin, H.,** "Neue Genossenschaften im Programm der integrierten Entwicklung in Bangladesh", *Zeitschrift für Ausländische Landwirtschaft*, Vol. 14, 1975, pp. 159-163

**Deutscher Bundestag,** Stenographisches Protokoll der 29. Sitzung des Ausschusses für wirtschaftliche Zusammenarbeit, Tagesordnung: Öffentliche Anhörung von Sachverständigen zum Thema "Armutsbekämpfung durch Selbsthilfe", 20 June 1988

**DSE,** "Selbsthilfeorganisation als Instrument der ländlichen Entwicklung", Seminarbericht, Berlin 1979

**Due, M.,** "Update on Financing Smallholders in Zimbabwe, Zambia, and Tanzania", *Savings and Development*, Vol. 7, 1983, No. 3, pp. 261-277

**Egger,** "Banking for the Rural Poor: Lessons from Some Innovative Savings and Credit Schemes", *International Labour Review*, Vol. 125, 1986, No. 4, pp. 447-462

**ESCAP,** *Case Studies on Strengthening Coordination between Non-Governmental Organizations and Government Agencies in Promoting Social Development*, United Nations, New York 1989

**Federal Ministry for Economic Cooperation (BMZ),** "Konzept zur Förderung der Ländlichen Entwicklung. Grundsätze für Planung und Durchführung von Vorhaben zur Entwicklung des Ländlichen Raumes", Bonn, 5 July 1988

-, *Querschnittsanalyse von Projekten der "ländlichen Regionalentwicklung*, Bonn 1990

-, *Sektorübergreifendes Konzept: Armutsbekämpfung durch Selbsthilfe*, Bonn 1989

**Fernando, E.,** "Informal Credit and Savings Organizations in Sri Lanka: The Cheetu System", *Savings and Development*, Vol. 10, 1986, No. 3, pp. 253-262

**Fischer, B., et al.,** "Sparkapitalbildung in Entwicklungsländern. Engpässe und Reformansätze", Forschungsberichte des BMZ, Vol. 78, Munich, Cologne, London 1986

**Fischer, W.E.,** "Eine Bank für die Ärmsten der Armen: Das "Grameen Bank Project" in Bangladesh", *Entwicklung und ländlicher Raum*, No. 3, 1983, pp. 13-16

**Friedrich Ebert Stiftung (ed.),** *Grundsätze für die Förderung von Selbsthilfeorganisationen*, Bonn 1979

**Geertz, C.,** "The Rotating Credit Association. A 'Middle Rung' in Development", *Economic Development and Cultural Change*, Vol. 11, 1962, pp. 241-163

**Geis, H.G.,** *Finanzierungskonzepte für den Selbsthilfebereich. Bank- und finanzwirtschaftliche Aspekte*, Schriftenreihe des Bundesministers für Jugend, Familien, Frauen und Gesundheit, Vol. 254, Stuttgart, Berlin / Cologne 1990

**German Development Institute (GDI),** *Grundbedürfnisorientierte ländliche Entwicklung*, Berlin 1980

**Ghosh, D.,** "Savings Behaviour in the Non-Monetized Sector and its Implications", *Savings and Development*, Vol. 10, 1986, No. 2, pp. 173-179

**Gotsch, C.H.**, "Credit Programs to Reach Small Farmers", *Development Digest*, Vol. 12, 1989, No. 2, pp. 165-178

**Goulet, D.**, "Participation in Development: New Avenues", *World Development*, Vol. 17, 1989, No. 2, pp. 165-178

**Green, R.H**, "Degradation of Rural Development: Development of Rural Degradation. Change and Peasants in Sub-Saharan Africa", IDS Discussion Paper No. 265, Sussex 1989

**Gsänger, H.**, *Perspektiven der Agrarreformpolitik Simbabwes im Lichte äthiopischer and kenianischer Erfahrungen*, GDI, Berlin 1985

**GTZ**, *Kaufkrafttransfer an die Ärmsten - Irrweg oder neuer Ansatz?*, Special publication of GTZ, No. 239, Eschborn 1989

-, *Ländliche Regionalentwicklung*, LRE kurzgefasst, GTZ publications, No. 207, Eschborn 1988

-, *Regional Rural Development, Guiding Principles*, Eschborn 1984

-, *Rural Finance, Guiding Principles*, Eschborn 1987

**Hayami, Y. / V. Ruttan**, *Agricultural Development*, Baltimore 1971

**Heidhues, F.**, "Zinskonditionen als Instrument ländlicher Entwicklungspolitik", *Entwicklung und ländlicher Raum*, No. 1, 1986, pp. 6-8

**Howell, J.**, "Government Services and Small Farmers", *Development Policy Review*, Vol. 3, 1985, pp. 89-102

**Informationsbrief Weltwirtschaft & Entwicklung**, "Armutsbekämpfung durch Selbsthilfe". Real existierender Zynismus, 21 September 1989

**Israel, A.**, *Institutional Development, Incentives to Performance*, Baltimore and London 1987

**Johnston, B.F. / P. Kilby**, *Agriculture and Structural Transformation. Economic Strategies in Late-Developing Countries*, New York, London, Toronto 1975

**Johnston, B. / J.W. Mellor**, "The Role of Agriculture in Economic Development", *American Economic Review*, Vol. 51, 1961, pp. 566-593

**Joint Working Group "Poverty Alleviation through Self-help"**, Ergebnisberichte AS 1/2 (savings and credit), AS 3 (fund for the promotion of the independent development of the poor), AS 4 (the land question in Latin America), AS 5 (informal sector), AS 6 (opportunities for and limits to poverty alleviation through formal self-help organizations), AS 7 (conceptual approaches to infrastructure measures), AS 8 (country-related development cooperation at governmental level for poverty alleviation through self-help), AS 9 (promotion of human abilities for poverty alleviation through self-help), Bonn 1989

**Jorgensen, D.W.**, "The Development of a Dual Economy", *Economic Journal*, Vol. 71, 1961, pp. 309-334

**Kirsch, O.C. / A. Benjakoc / L. Schujmann,** *The Role of Self-Help Groups in Rural Development Projects*, Publications of the Research Centre for International Agrarian Development, Saarbrücken, Fort Lauderdale 1980

-, "Finanzierungsinstruments und Selbsthilfeeinrichtungen zur Förderung ärmerer Zielgruppen", FIA-Berichte 84/1, Heidelberg 1984

**Kirsch, O.C. / P.G. Armbruster / G. Kochendörfer-Lucius,** "Selbsthilfeeinrichtungen in der Dritten Welt", Forschungsberichte des BMZ, Vol. 49, Munich, Cologne, London 1983

**Kropp, E.,** "Ländliches Finanzwesen. Bestandteil von Vorhaben der Technischen Zusammenarbeit zur Ländlichen Entwicklung", *Entwicklung und ländlicher Raum*, No. 1, 1986, pp. 23-26

**Kropp, E. / M.T. Marx / B.R. Quinonesd / H.D. Seibel (eds),** *Linking Self-Help Groups and Banks in Developing Countries*, GTZ/ APRACA, Eschborn 1989

**Kumar, J.,** *Integrated Rural Development. Perspectives and Prospects (1952 - 1982)*, Delhi 1987

**Kumar, K.,** "A.I.D.'s Experience with Integrated Rural Development Projects", A.I.D. Program Evaluation Report No. 19, Washington 1987

**Lachenmann, G.,** "Soziale Bewegungen als gesellschaftliche Kraft im Demokratisierungsprozess in Afrika?", *Afrika Spektrum*, Vol. 26, 1991, No. 1, pp. 73-97

**Lele, U.,** *The Design of Rural Development: Lessons from Africa*, Baltimore / London 1975

**Lipton, M.,** "Agricultural Finance and Rural Credit in Poor Countries", *World Development*, Vol. 4, 1976, No. 7, pp. 543-553

-, "Agriculture, Rural People, the State and the Surplus in Some Asian Countries. Thoughts on Some Implications of Three Recent Approaches in Social Science", *World Development*, Vol. 17, 1989, No. 10, pp. 1553-1571

**Long, M.,** "Conditions for Sucess in Small Farmer Credit Programs", *Development Digest*, Vol. 12, 1974, No. 2, pp. 47-53

**Manig, W.,** *Integrierte rurale Entwicklung*, Kiel 1985

**McCullogh, J.S. / R.W. Johnson,** "Analysing Decentralization Policies in Developing Countries. A Political-Economy Framework", *Development and Change*, Vol. 20, 1989, pp. 57-87

**Mellor, J.W.,** *The New Economies of Growth*, Ithaca 1976

**Meyer, R.L. / C. Gonzalez-Vega,** "Rural Deposit Mobilization in Developing Countries", *Entwicklung und ländlicher Raum*, No. 1, 1986, pp. 9-11

**Mittendorf, H.J.**, "Promotion of Viable Rural Financial Systems for Agricultural Development", *Zeitschrift für Ausländische Landwirtschaft*, Vol. 26, 1987, No. 1, pp. 6-27

**Mosely, P.**, "Crop and Livestock Insurance Schemes in Less Developed Countries. Some Issues of Design", *Savings and Development*, Vol. 13, 1989, No. 1, pp. 5-20

**- / R.P. Dahal**, "Lending to the Poorest. Early Lessons from the Small Farmers Development Programme in Nepal", *Development Policy Review*, Vol. 3, 1985, pp. 193-207

**Müller-Glodde, U. / G. Urban**, "GTZ-Orientierungsrahmen "Fonds zur Förderung von Selbsthilfe". Beschreibung eines Instruments", unpublished draft, Eschborn 1990

**- / G. Bedard / H. Heussen (eds)**, "Selbsthilfeförderung durch Sparkassen." Bericht über ein Dialogprogramm veranstaltet von DSE/DSGV/BMZ, Berlin 1987

**OECD**, *Voluntary Aid for Development. The Role of Non-Governmental Organisations*, Paris 1988

**Padmanabhan, K.P.**, "Credit Planning for Rural Development", *Savings and Development*, Vol. 6, 1982, No. 2, pp. 197-209

-, "Giving Credit Where Due. The Record of India's Rural Banking System", *Ceres*, 1987, No. 115, pp. 33-37

**Panday, D.R.**, "From Dependence to Self-Reliance. The Future of Technical Cooperation", *Development & Cooperation*, No. 6, 1989, pp. 6-9

**Pischke, J.D. von / J. Rouse**, "Selected Successful Experiences in Agricultural Credit and Rural Finance in Africa", *Savings and Development*, Vol. 7, 1983, No. 1, pp. 21-44

**Platteau, J.P. / A. Abraham**, "Credit as an Insurance Mechanism in the Backward Rural Areas of Less Developed Countries", *Savings and Development*, Vol. 8, 1984, No. 2, pp. 115-133

**Ranis, G. / J.C.H. Fei**, "A Theory of Economic Development", *American Economic Review*, Vol. 51, 1961, pp. 252-273

**Rauch, T.**, "Multisektorale ländliche Entwicklungsprogramme - ein gescheiterter Ansatz in der Technischen Zusammenarbeit?", unpublished manuscript, Berlin 1989

-, *LRE-aktuell*, "Strategieelemente für eine Umsetzung des LRE-Konzeptes unter veränderten Rahmenbedingungen", study commissioned by GTZ, Eschborn 1991

**Ray, J.K.**, *Organising Villagers for Self-Reliance. A Study of Deedar in Bangladesh*, Comilla 1983

**Rondinelli, D.A.**, "Decentralization, Territorial Power and the State. A Critical Response", *Development and Change*, Vol. 21, 1990, pp. 491-500

**Ruthenberg, H.**, "Is Integrated Rural Development a Suitable Approach to Attack Rural Poverty?", *Quarterly Journal of International Agriculture*, Vol. 20, 1981, No. 1

-, "Rural Development - eine Sackgasse?", *Zeitschrift für Ausländische Landwirtschaft*, Vol. 14, 1975, pp. 3 f.

**Salmen, L.**, "Institutional Dimensions of Poverty Reduction", World Bank, PRE Working Papers, WPS 411, Washington 1990

**Samoff, J.**, "Decentralization: The Politics of Intervention", *Development and Change*, Vol. 21, 1990, pp. 513-530

**Sanderatne, N.**, "The Political Economy of Small Farmer Loan Delinquency", *Savings and Development*, Vol. 10, 1986, No. 4, pp. 343-353

**Schaefer-Kehnert, W.**, "Staatlicher Agrarkredit. Politisches Steuerinstrument oder Hemmschuh ländlicher Entwicklung", *Entwicklung und ländlicher Raum*, No. 1, 1986, pp. 20-22

**Schoop, W. / H. Knauss**, "Armutsbekämpfung durch Selbsthilfe - Wovon ist die Rede?", evaluation of statements made at the public hearing held by the Committee on Economic Cooperation on 20 June 1988, unpublished working paper, year not given

**Schubert, B. / G. Balzer**, "Soziale Sicherungssysteme in Entwicklungsländern. Transfers als sozialpolitischer Ansatz zur Bekämpfung überlebensgefährdender Armut", Sonderpublikation der GTZ, No. 246, Eschborn 1990

**Schultz, T.W.**, *Transforming Traditional Agriculture*, New Haven 1964

**Seibel, H.D.**, "Duale Finanzmärkte in Afrika. Modernisierung traditioneller oder Traditionalisierung moderner Finanzinstitutionen", *Entwicklung und ländlicher Raum*, No. 1, 1986, pp. 14-16

**Seidl, C.**, "Poverty Measurement: A Survey", in: D. Bös, M. Rose, C. Seidl (eds), *Welfare and Efficiency in Public Economics*, Berlin, Heidelberg, New York 1988

**Sen. A. / J. Drèze**, *Hunger and Public Action*, London 1989

**Senghaas, D.**, "Integrierte ländliche Entwicklung", *epd - Entwicklungspolitik*, No. 1, 1981, pp. 23-25

**Siegert, H.**, *Ländliche Entwicklung und Bankpolitik in Entwicklungsländern. Zur Problematik der Kreditversorgung kleinbäuerlicher Betriebe aus systemtheoretischer Sicht*, Frankfurt, Bern, New York, Paris 1987

**Sing, I.**, *The Great Ascent. The Rural Poor in South Asia*, Baltimore / London 1990

**Slater, D.**, "Territorial Power and the Peripheral State: The Issue of Decentralization", Vol. 20, 1989, pp. 501-531

-, "Debating Decentralization - A Reply to Rondinelli", *Development and Change*, Vol. 21, 1990, pp. 501-512

**Stamm**, V., "Some Sociological Remarks on Rural Development", *Zeitschrift für Ausländische Landwirtschaft*, Vol. 29, 1990, No. 2, pp. 169-172

**Steigerwald, V.**, "Systemorientierung in der integrierten ländlichen Entwicklung", *Studien zur Integrierten Ländlichen Entwicklung*, Vol. 31, Hamburg 1989

**Steiner, A.**, "Soziale Sicherungssysteme in Entwicklungsländern, Strategische und konzeptionelle Aufgaben für die Entwicklung eines neuen Leistungsangebotes der GTZ", unpublished, Sept. 1990

**Stockhausen, J. von**, "Agrarkreditpolitik in Entwicklungsländern", *Berichte über Landwirtschaft*, , Vol. 61, 1983, pp. 609-632

-, "Ländliche Finanzmarktpolitik im Rahmen hoher gesamtwirtschaftlicher Verschuldung", *Entwicklung und ländlicher Raum*, No. 1, 1986, pp. 3-5

**Thun, F. von / G.J. Ullrich (eds)**, "Fighting Rural Poverty through Self-help", Report on a Conference organised by DSE, Feldafing 1985

**Tinnermeier, R.**, "Technology, Profit, and Agricultural Credit, *Development Digest*, Vol. 12, 1974, No. 2, pp. 54-60

**UN**, *Interagency Committee on Integrated Rural Development for Asia and the Pacific, Integrated Rural Development in Asia and the Pacific. A Framework for Action for the 1990s*, Bangkok 1989

**UNDP**, *Human Development Report 1990*, New York 1990

**Waller, P.P.**, *Grundbedürfnisorientierte Regionalentwicklung*, GDI, Berlin 1984

**Wilson, P.A.**, "Linking Decentralization and Regional Development Planning. The IRD Project in Peru", *American Planning Association Journal*, Summer 1987, pp. 348-357

**Wissenschaftlicher Beirat beim BMZ**, *Möglichkeiten und Grenzen der Selbsthilfe im Rahmen einer armutsorientierten Entwicklungspolitik*, Bonn 1989

**World Bank**, *Poverty Reduction and Bank Operations*, Washington 1990

-, "Rural Development", Sector Policy Paper, Washington 1975

-, *Rural Development. World Bank Experience, 1965 - 1986*, Washington 1989

-, *Sub-Saharan Africa. From Crisis to sustainable Growth*, Washington 1989

-, *World Development Report 1990 (Poverty)*, Washington 1990

**Yudelman, M.**, "Agricultural Development in the Third World: The World Bank Experience, Social Strategies", Forschungsberichte Soziologisches Seminar der Universität Basel, Vol. 1, 1986, No. 3

-, "Die Rolle der Landwirtschaft bei Projekten der integrierten ländlichen Entwicklung. Die Erfahrungen der Weltbank", in: Th. Dams (ed), *Integrierte ländliche Entwicklung*, München, Mainz 1980, pp. 120-138

www.ingramcontent.com/pod-product-compliance
Ingram Content Group UK Ltd.
Pitfield, Milton Keynes, MK11 3LW, UK
UKHW041840280225
455677UK00010B/258